International Conference on Materials Applications and Engineering (ICMAE 2017)

MATEC Web of Conferences Volume 142 (2018)

Qingdao, China
25 - 27 August 2017

Editor:

Huining Xiao

ISBN: 978-1-5108-5579-3

Printed from e-media with permission by:

Curran Associates, Inc.
57 Morehouse Lane
Red Hook, NY 12571

proceedings.com

Some format issues inherent in the e-media version may also appear in this print version.

Printed by Curran Associates, Inc. (2018)

For additional information, please contact EDP Sciences – Web of Conferences at the address below.

EDP Sciences – Web of Conferences
17, Avenue du Hoggar
Parc d'Activité de Courtabœuf
BP 112
F-91944 Les Ulis Cedex A
France

Phone: +33 (0) 1 69 18 75 75
Fax: +33 (0) 1 69 28 84 91

contact-edps@webofconferences.org

Additional copies of this publication are available from:

Curran Associates, Inc.
57 Morehouse Lane
Red Hook, NY 12571 USA
Phone: 845-758-0400
Fax: 845-758-2633
Email: curran@proceedings.com
Web: www.proceedings.com

TABLE OF CONTENTS

Preface

This is the proceedings of selected papers presented at the International Conference on Materials Applications and Engineering 2017 (ICMAE2017) held in Qingdao, China, on Aug. 25-27, 2017, which was organized by Shandong University of Science and Technology. The main objective of ICMAE2017 is to bring together researchers from different scientific institutions and universities to present and discuss latest developments and trends in kind of materials science.

ICMAE is an important serial conference of Shandong University of Science and Technology and has become a popular conference for a wide range of topics as Chemical Materials and Chemical Syntheses; Building Materials and Properties; Metals, Alloys and Coatings etc. Approximately, the organizers received 70 technical papers and more than 45 participants attended the conference. After the peer-review processes and by recommendations of technical committees, these submitted papers were selected on the basis of originality, significance, and clarity for the purpose of the conference; finally 35 papers are published in the present proceedings.

The organizing scientific committee of this conference hopes that all participants enjoyed the conference, got new impressions and had a good time in Qingdao. We would like to give the thanks to the conference chairs and all the paper reviewers, especially MATEC Web of Conferences who worked so hard to provide the help for the conference.

Look forward to the coming of ICMAE in 2018.

Editors
Prof. Huining Xiao

Committees

Honorary Chairs

Prof. Min-Shiun Chao, National Chin-Yi University of Technology, Taiwan
Prof. Chen Jingsheng, Department of Materials Engineering, National University of Singapore, Singapore
Prof. Hong-Wei Ma, Xi'an University of Science & Technology, China
Prof. Elena I. Korotkova, Department of Physical and Analytical Chemistry of TPU, Tomsk, Russia

General Chairs

Prof. Huining Xiao, Department of Chemical Engineering, University of New Brunswick, Canada
Prof. Hongfei Sun, School of Materials, Shandong University of Science and Technology, China

Organizing Committees

Prof. Shenhao Han, School of physics, Shandong University, China
Prof. Jeng-Da Chai, Department of Physics, National Taiwan University, Chinese Taipei

Technical Program Committees

Prof. Huan-Tsung Chang, Analytical Chemistry Nanotechnology and Green Chemistry, National Taiwan University, Chinese Taipei
Prof. Osman Adiguzel, Department of Physics, Firat University, Turkey
Prof. Daniel Choi, Mechanical and Materials Engineering Department, Masdar Institute, UAE
Dr. Joo Eun Chung, Institute of Bioengineering and Nanotechnology (IBN), Singapore
Prof. Manish Kumar, Materials School of Advanced Materials Science and Engineering, Sungkyunkwan University, Korea
Prof. Kienwen Sun, Center for Nano Science and Technology, National Chiao Tung University, Taiwan
Dr. Sushanta Kumar Panigrahi, Materials Engineering, Indian Institute of Technology Madras, India
Prof. Chong Soo Lee, Molecular Biology, Pohang University of Science and Technology, South Korea
Dr. Xiaowen Yuan, School of Engineering and Advanced Technology, Massey University, New Zealand
Prof. Jingwei Zhao, School of Mechanical, Materials & Mechatronic Engineering, University of Wollongong, Australia
Prof. Tomasz Wejrzanowski, Faculty of Materials Science and Engineering, Warsaw University of Technology, Poland
Prof. Andrey N. Zagoruiko, Sc. Chemistry, Leading research engineer in Boreskov Institute of Catalysis SB RAS, Novosibirsk, Russia
Dr. Li Chang, State Key Laboratory of Food Science and Technology, Nanchang University, China

Statement of Peer review

In submitting conference proceedings to MATEC Web of Conferences, I certify to the Publisher that I adhere to the **Policy on Publishing Integrity** of the journal in order to safeguard good scientific practice in publishing.

1. All articles have been subjected to peer review administered by the proceedings editors.
2. Reviews have been conducted by expert referees, who have been requested to provide unbiased and constructive comments aimed, whenever possible, at improving the work.
3. Proceedings editors have taken all reasonable steps to ensure the quality of the materials they publish and their decision to accept or reject a paper for publication has been based only on the merits of the work and the relevance to the journal.

Title, date and place of the conference

International Conference on Materials Applications and Engineering 2017(ICMAE2017)
Aug. 25-27, 2017
Qingdao, China

Proceedings editor(s):

Prof. Huining Xiao

Date and editor's signature

17, avenue du Hoggar · PA de Courtaboeuf · BP 112 · 91944 Les Ulis Cedex A (France)
Tél. 33 (0)1 69 18 75 75 · Fax 33(0)1 69 07 45 17 · www.edpsciences.org

Control Materials of Hydrogen Concentration in Micro Environment and Their Summary of Application

Min Zhu[1] , *Cuihua* Lu[2 *] , and *Hongqing* Zhou[2]

[1]Naval Nniversity of Engineering, Wuhan, Hubei Province, China
[2]Naval Aviation University, Yantai, Shandong Province, China

Abstract. Some materials will be hydrogenated to corrosion, such as metal materials, organic materials, polymer materials and so on, when the concentration of hydrogen in a mixed gas is greater than 4% in a closed space of micro environment. Their performances will be seriously affected. So removing or reducing the concentration of hydrogen in a micro environment can fundamentally solve the problem of hydrogenation corrosion. Metallic hydrogen removing agent and organic hydrogen removing agent are the main research directions of controlling materials of hydrogen concentration in micro environment. Current research status in the world are comprehensively reviewed from these two aspects. Features and application scopes of these two kinds of materials are pointed out separately, which provide a theoretical reference for the prevention of accidents caused by hydrogenation corrosion.

1 Introduction

If a complex atmosphere hydrogen removing material meets these 6 principles[1] proposed by R.L. Courtney. In the report, it can ensure the elimination of hydrogen without harmful side effects. Metallic hydrogen absorbing materials and organic hydrogen absorbing agent can meet these 6 principles.

The mechanism of metallic hydrogen absorbing material is that the surface of conventional metallic hydrogen-absorbing material whose hydrogen absorption properties meet requirements is covered with a layer of anti-poison material, which can be isolated from the poisonous gas and can avoid the impact on the hydrogen absorption properties of the internal material which is easy to be poisoned.

The mechanism of organic hydrogen absorbing agent is as the formula (1) and formula (2):

$$R^{'}-C\equiv C-R+Cat\rightarrow R^{'}-\underset{\underset{Cat^{*}}{|}}{C\equiv} C-R \quad (1)$$

* Corresponding author: llu1978@163.com

$$R'-C\equiv C-R+Cat\rightarrow R'-\underset{\underset{Cat^*}{|}}{C}\equiv C-R \quad (2)$$

2 Research status and development trend of metallic hydrogen-absorbing materials

Metallic hydrogen-absorbing materials mainly include ionic hydrogen absorbing material, alloy hydrogen absorbing material, carbon and Nano hydrogen absorbing material, etc.

2.1 Ionic hydrogen absorbing material

Ionic hydrogen absorbing material is an early hydrogen absorbing material. Study on synthesis and application of this kind of material has began in the 50's. Early applications are directly used as reducing agents.

Bogdanovic found that the hydrogen storage capacity of modified NaAlH4 with Ti can reach 3.1%-4.4% (mass fraction) and its cycle stability is also good.

Ionic hydride is generally used as a reducing agent and to supply a small amount of hydrogen, especially in underwater applications. Its unique advantages is whose gas product is only hydrogen, but ionic hydride is not optimistic on the prospects for large-scale applications.

2.2 Alloy hydrogen absorbing material

Alloy hydrogen absorbing material is a metal compound which can absorb, store and release hydrogen in large quantities under a certain temperature and hydrogen pressure. Alloy hydrogen absorbing material is the most widely used hydrogen absorbing material at present, because of its large hydrogen storage capacity, no pollution, safe and reliable, and relatively mature preparative technique. Alloy hydrogen absorbing material is mainly divided into magnesium series (A_2B type), rare earth series (AB_5 type), titanium series (AB type) and zirconium series(AB_2 type).

2.2.1 Mg-based hydrogen absorbing materials

Mg-based hydrogen absorbing materials are represented by Mg_2Ni. The American brooke haven national laboratory first smelted into Mg_2Ni with magnesium and nickel in 1968. Preliminary research on Mg- $LiBH_4$ system is carried out by Anderson et al[2], who explored a way to improve the hydrogen absorption properties of magnesium without activation. They found that $LiBH_4$ could effectively improve the hydrogen absorbing kinetics of magnesium. Hydrogen absorbing mass fraction of Mg-$LiBH_4$ can reach 6.5% in 20min at 300 ℃, which laid the foundation for the Mg based hydrogen absorbing material to be used as the hydrogen source of fuel-cell vehicles. Nano composite hydrogen absorbing materials have been made by Balooch etc. through high-energy ball-milling magnesium powder and crystalline or amorphous ZrNiCr and $ZrNi_{1.6}Cr_{0.4}$ powder. [3] Stone etc. [4] have made Mg-based hydrogen storage materials containing carbon nanotubes (Mg-3Ni-2MnO_2-0.25CNTs) by mechanical alloying and using H_2 as a protective atmosphere, whose hydrogen storage mass fraction is 7.0%. The hydrogen absorption process of the material will be completed in 100s, and its hydrogen desorption process will be completed in 600s at 0.1MPa and the about 280℃. Mroz[5] made Nano composite hydrogen absorbing material(Mg-3Ni-2MnO_2) by ball milling. It has a very good performance in circulating hydrogen storage, whose maximum hydrogen storage capacity is 6.5% and which can absorb 6.37% of H_2 in 50s and at 200℃. After 60 cycles, the decrease in hydrogen storage is less than 0.1%, but the time required for the absorption and desorption of hydrogen is increased with the increase of the number of cycles. Colmenares etc. [6] made $Mg_{100-x}Ni_x$(x=5,11.3,20,25) through the hydrogen burning-mechanical ball milling method. Alloy can be crushed and refined by ball milling process. Hydrogen desorption temperature of alloy which made through hydrogen burning-mechanical ball milling method is much lower than that of alloy which made through hydrogen burning method. The maximum hydrogen storage capacity of $Mg_{80}Ni_{20}$ is 2.77 % in 600s and at 313K. The maximum hydrogen storage capacity of $Mg_{95}Ni_5$ is 4.88 % at 313K, and The maximum neon storage capacity is 5.41 % at 473K. The hydrogen storage capacity of Mg-based hydrogen absorbing materials is big, but their kinetics performance and cycle life in alkaline solution are poor, then their properties need to be improved in these two areas.

2.2.2 Rare earth series hydrogen absorbing materials

Rare earth series hydrogen absorbing materials have $CaCu_5$-type hexagonal structure. Their representative is $LaNi_5$. Rare earth alloy has the advantages that its hydrogen absorbing rate is fast and it is easy to activate and so on, but it is not satisfactory in terms of performance and cost hydrogen storage, etc. Its performance can be improved by using the multi-alloy, which uses Mm (mixed rare earth), Ca, Ti and other to replace parts of La, and uses Co, Al, Mn, Sn and other to replace parts of Ni. Powell's [7]research from the perspective of electronic shows that the performance of the alloy can greatly affect by using Cu and other 3D series metal or Al, Sn and other S-P series metal to replace Ni in $LaNi_5$. Its hydrogen storage capacity can be increased, but the stability will be affected. Balooch[8] etc. prepared Nano $LaNi_{4.2}Al_{0.8}$ alloy through mechanical alloying method, whose hydrogen storage performance has been significantly improved. Research made by Yuan Xianxia et al. about $MINi_{3.75}Co_{0.65}Mn_{0.4}Al_{0.2}$ alloy shows that with the increase of temperature the hydrogen absorption and desorption rate of the hydrogen storage alloy is accelerated in the range of 253K-358K. Hai Zhenwei[9] etc prepared $LaNi_5$ series alloy powder, whose has the advantages of accurate chemical composition, high purity, small particle size and narrow distribution, etc. Compared with molten alloy it has better activation property, which shows that rare earth alloy has potential research value and application value.

2.2.3 Titanium series hydrogen absorbing materials

FeTi is the representative of titanium series hydrogen absorbing materials. Its advantage is that the hydrogen temperature is low and the price is moderate, but it is not easy to be activated, and easy to be poisoned by impurities such as H_2O and O_2, serious lag and unstable life. Now Ni and other metal parts are used to replace Fe to form ternary alloy in order to achieve room temperature activation. The results show that the activation properties of the alloy can be improved obviously by using Mn, Cr, Zr and Ni etc transition element to replace partial Fe of FeTi alloy. When the purity of H_2 is more than 99.5%, its cycle life is more than 26000 times. In addition, the activation property of FeTi alloy can be improved by mechanical milling and surface treatment with acid, alkali and other chemical reagents. At present, this kind of material has been widely used in Germany, the United States and other developed countries. Japan Metal Materials Technology Research Institute successfully developed the titanium- iron-oxide hydrogen storage system, which has the advantages of large amount of hydrogen absorption, fast hydrogenation, easy activation and so on.

2.2.4 Zirconium series hydrogen absorbing materials

$ZrMn_2$ is the representative of zirconium series hydrogen absorbing materials. Zr based alloys have a rich phase structure, such as C14, C15 and C36 Laves phase and a variety of B.C.C (solid solution). The mechanism and synergistic effect of each phase are currently being studied. The alloy has the advantages of high hydrogen absorption and desorption capacity, easy activation, small thermal effect (2-3 times smaller than LaNi5 and other materials) and so on. The properties of the multivariate Zr based hydrogen storage alloys are better which made by partial subtitution of Zr with Ti and partial subtitution of V,Cr,Mn with Fe,Co,Ni.

2.2.5 Composite hydrogen absorbing alloy materials

In recent years, the research of composite hydrogen absorbing alloy materials has made a breakthrough. Nguyen[10] et al. found that the ball milled Mg_2Ni-Ni complex (70%Ni) may be a non crystalline structure. Compared with Mg_2Ni, its hydrogen absorption rate is faster and its hydrogen absorption capacity is higher (up to 4%), but the stability is poor. $Mg_{1.0}Ni_{1.0}V_{1.0}$ alloy prepared through ball milling by VSalazar[11] et al. Its hydrogen absorption capacity is 2.3%, at 298K and 3MPa. The most prominent feature of this alloy is that it can store hydrogen at room temperature. Hydrogen storage capacity of $La_2Mg_{17-x}LaNi_5$ complex developed by Carroll[12] et al. can be up to 5.3%, but its activation temperature is high. Absorbing and disrobing hydrogen capacity of Mg- 30%$LaNi_5$ complex can be up to 5% at 573K, which is made by Kaszuba[13] et al. through high-temperature sintering. Absorbing and disrobing hydrogen capacity of Mg-30%$LaNi_5$ and MgH_2-30%$LaNi_5$ can be up to 5% at 573K, and its dehydriding kinetics has been improved significantly, which are made by Holtz[14] et al. through ball-milling method. The hydrogen storage capacity of Mg-30%$LaNi_{2.28}$ complex is 5.4% at 553K and 3MPa, which is made by Rodrigo[15] et al. through mechanical alloyin. The hydrogen storage capacity of $La_{1.5}Ni_{0.5}Mg_{17}$ can be up to 7.01% at 573K, which is made by Giorgi[16] et al. using the same method. Even at 373K its hydrogen storage capacity can be up to 4.03% and the speed of hydrogen absorption and desorption is fast. Zhu et al. reported that the electrochemical properties and hydrogenation behavior of $Mg_2Ni/MmNi_{5-x}$(CoAlMn) composites are better than single-phase alloy. Research made by Cotton[17] et al. shows that the hydrogen absorption capacity of Mg-30%$TiMn_{1.5}$ nanocomposites can be up to 4.4% at 523K~573K and its saturated hydrogen storage capacity can also be up to 2.7% at 373K. Henrie et al.[18] reported that hydrogenation properties and hydrogen storage efficiency of new nanocomposites have been significantly improved which is made through adding organic solvent(tetrahydrofuran, cyclohexane etc.) in magnesium and graphite by mechanical grinding.

2.3 Carbon and nano hydrogen absorbing materials

In recent years, due to the research upsurge of nano materials, carbon and nano-carbon materials for hydrogen storage have become a hot research topic. Carbon hydrogen absorbing materials mainly include high specific surface Activated carbon(AC), graphite nanofibers (GNF), carbon nanofibers(CNF) and carbon nano-tube(CNT).

2.3.1 High specific surface activated carbon

Carpetis[19] first proposed that hydrogen can be stored in activated carbon by adsorption under the condition of low temperature. They first mentioned in their paper that low temperature adsorbent can be applied to the large hydrogen storage system, and pointed out that hydrogen's storage volume density in the active carbon is quite to liquid hydrogen's volume density. Zhou Li et al. used super activated carbon in hydrogen storage whose specific surface area is 3000 m^2/g and micropore volume is 15mL/g(on the basis of CO_2 adsorption). The hydrogen absorption capacity is 5% at 77K and 3Mpa, but with the increasing of temperature the hydrogen storage capacity is getting lower and lower. In addition, Zan Liang et al. pointed out that surface per unit mass activated carbon is larger, the greater amount of hydrogen can be adsorbed, which is certainly true especially in a high pressure. Lewis's[20] work focused on the study on mechanism of hydrogen absorption on activated carbon at the end of 1980s and early 90's.Best results obtained is that hydrogen adsorption capacity is 4.8% at 87Kand 59atm through surface treatment of activated carbon. Heung et al.[21] studied that effect of gas impurities in hydrogen on the adsorption properties of activated carbon, pointed out that if super activated carbon was used as an adsorbent, hydrogen adsorption amount decreases no more than 30% which is caused by nitrogen impurities in hydrogen whose concentration is in the range of 500ppm. Tamm et al.[22] tested the hydrogen storage properties of a series of carbon sorbent at 300K, 190K and 80K, such as activated carbon, activated carbon fiber, carbon black, carbon airgel, carbon molecular sieve. The results showed that only one test adsorbent's storage capacity can be increased in ten kinds of test adsorbents in about 200bar pressure and there was only a small amount of increment at 190K and 300K, no increment in 80K. Overall, there is no report on activated carbon adsorbent at room temperature has better hydrogen storage properties.

2.3.2 Graphite nanofibers

Statement made by Devillers et al.[23] that graphite nanofibers have high hydrogen adsorption storage capacity sparked people's interest in this material. These materials are a mixture of ethylene, hydrogen and carbon monoxide. Through Choosing different catalyst, 3 different structures of graphite nanofiber scan be formed: tubular, flat-plate, fishbone shaped. The hydrogen storage capacity of three structures of graphite nanofiber are respectively 11 wt %, 53 wt % and 67 wt % at room temperature and at 11.35Mpa, which were measured by Woyke et al.[24] But the conclusion was too high which the hydrogen adsorption capacity of fishbone shaped graphite nanofiber was 67%, and then it was denied by other research groups. The validation results showed that the maximum amount of hydrogen storage at room temperature of graphite nanofiber is only between 0.08%-1.5%. Therefore, Penzhorn et al.[25] believed that high density hydrogen storage of graphite nanofibers is no hope.

2.3.3 Carbon nanofibers

Carbon nanofibers(CNF) is the carbon fiber that its structure is solid and its fiber diameter is less than 1000nm. Vapor-grown carbon nanofibers whose diameter is about 80nm made by Bonizzoni et al. [26] through floating catalyst method. Its hydrogen absorption capacity can be up to 12 wt% at room temperature and 11Mpa without treatment. Hydrogen absorption capacity of carbon nanofibers whose diameter is between 90nm-125nm can be up to about 10 wt% after ethanol dispersion and hydrochloric acid. Doni et al. [27] reported that the highest hydrogen storage capacity of carbon nanofibers is 10wt% after making carbon nanofiber to open and removing impurity gas adsorbed on the surface of the fiber. Perevezentsev et al. [28] have measured that carbon nanofibers can store about 10wt% hydrogen too. Srivatava et al. [29] found that carbon nanofibers daubed with lithium and potassium could adsorb 14wt% aydrogen at 473K and 20wt% at 673K under environmental pressure, but later they guessed the adsorption capacity measured might contain water vapor. Chambers et al.[30] developed carbon nanofibers whose average diameter was about 100nm. Its hydrogen adsorption capacity is about l0wt% at 8-11Mpa and room temperature. These results still have not been proved by other people theoretically or experimentally. The computer simulation of nanofiber hydrogen adsorption can not explain these phenomena. The current computer simulation focuses on physical adsorption of hydrogen storage, and Diederich[31] proved even if the chemical adsorption exists, it is impossible to achieve the adsorption amount.

2.3.4 Carbon nanotubes

Since the discovery of carbon nanotubes in the accident in 1991 by Iijima, it quickly became a hot research topic in many high-tech fields due to its unique one-dimensional tubal structure and peculiar mechanical, electronic, properties etc. Carbon nanotubes can be divided into single-walled carbon nanotube (SWNT)and multi-walled carbon nanotube(MWNT). Sane et al. [32] first studied hydrogen adsorption performance of single-walled carbon nanotube without purification whose mass fraction is only 0.1-0.2 wt% by using TPD method. Estimated from the relevant results, the pure SWNT can absorb hydrogen at about 5-l0wt% at room temperature and 40kPa, which is ten times the amount of activated carbon hydrogen storage and its adsorption heat is 19.6kJ·mol^{-1}. Ichimura et al.[33] have carried on the analysis to the interaction of hydrogen which is in SWNT. In their experiments, analyzable spectrum was tested through the thermal desorption spectrum and the main reversible desorption is in the range of 77K-320K. At about 90K, the peak value of the activation energy was assumed to be the first sequence of desorption, which corresponded to the physical surface adsorption of single-walled carbon nanotubes. Janberg et al. [34] purified single-walled carbon nanotubes and studied the connection between the hydrogen storage of the single-walled carbon nanotube and its surface area. They found that the hydrogen storage capacity of SWNT is 8wt%, which was far more than traditional hydrogen storage system, but the hydrogen storage capacity of MWNT is less capable.

3 Research status and development trend of organic hydrogen absorbing agent

America's Sandia National Laboratories has used phenol, potassium hydroxide, chloro propyne etc. to form acetylenic compounds: phenyl propargyl ether(PPE) and dimerization of phenyl propargyl ether(DPPE) in the early 70's. Meanwhile the researchers used palladium black, calcium carbonate, asbestos or the 5% palladium on alumina and PPE or DPPE for the preparation of hydrogen absorbing agent. Their research work proved that unsaturated organic compounds containing one or more of triple bonds had a superior absorption capacity in a closed system. Heterogeneous catalysis has be also proved that it was a good way to activate the hydrogenation reaction. The application scope of early organic hydrogen absorbing agent is limited because of their low melting point, such as DPB(1,4-diphenyl butadiyne)and DPPE. C A Colmenares has tested the equilibrium vapor pressures of DPB and DPPE. In the range of 23~75℃, the vapor pressure of DPB vary in the range of $6.6\times10^{-7}\sim6.0\times10^{-4}$ Torr and that of DPPE vary in the range of $3.5\times10^{-7}\sim 8.4\times10^{-7}$Torr. G L Powell studied hydrogenation kinetics of DPB. The initial rate of DPB hydrogenation is slow and rate increases with the deepening of the reaction degree. Before reaching the theoretical maximum degree of reaction it keeps the reaction rate higher than the initial rate. The higher reaction rate of DPB in the high degree of reaction may be caused by eutectic which made by reactant mixture.

In the 1990s, Sheppod et al. tested many potential hydrogen compounds, and found DEB(1,4-bis(phenylethynyl)benzene) had the best performance.

DEB's chemical structure is shown as the formula (3):

(3)

DEB is non-toxic, non-mutated crystalline solid. The melting point of DEB is 179 ℃ and all products of hydrogenation melt at 87℃. The standard component of DEB hydrogen absorbing agent is a mixture of 75% DEB and 25% carbon catalyst (with 5% Pd on carbon). At 70 degrees and under the environment of N_2, the hydrogen absorbing agent can be stable for 18 months without the presence of hydrogen. Because DEB contains two triple bonds, 1mol DEB can react with 4mol hydrogen. DEB and hydrogen reaction is fast, exothermic, irreversible. Hydrogen storage capacity per gram DEB is in 240 ~ 330cm^3. The reaction is close to the stoichiometric, and reachs > 90% of the theoretical level. If hydrogen is injected at the rate of $10^{-3}cm^3/sec$ in nitrogen environment, the hydrogen concentration can be kept under 5ppm until the reaction degree of DEB reachs > 90% of the theoretical level. The reaction rate of DEB and hydrogen is faster with the increase of the temperature. The study also found that CO and several chlorinated volatile organic compounds inhibited the reaction of DEB and hydrogen. M. Balooch et al. was also measured pure DEB, the mixture of DEB and carbon supported palladium (DEB-Pd/C), vapor pressure of hydrogenated DEB-Pd/C from room temperature to the melting point (179 ℃). The results showed that vapor pressure of hydrogenated DEB-Pd/C was the highest. DEB-Pd/C's vapor pressure was higher than the predictive value through the ideal solution model, which showed that there were positive interaction when DEB and carbon supported palladium form DEB-Pd/C.

4 Conclusion

Metallic hydrogen removing agent and organic hydrogen removing agent are the main research directions of controlling materials of hydrogen concentration in micro environment. Current research status in the world are comprehensively reviewed from these two aspects.Metallic hydrogen absorbing materials mainly include ionic hydrogen absorbing material, alloy hydrogen absorbing material , carbon and nano hydrogen absorbing material, etc.Alloy hydrogen absorbing material is the most widely used hydrogen absorbing material at present, because of its large hydrogen storage capacity, no pollution, safe and reliable, and relatively mature preparative technique.In recent years, carbon and nano-carbon material for hydrogen storage is a hot research topic. Research found that the hydrogen storage capacity of SWNT is far more than traditional hydrogen storage system.The application scope of early organic hydrogen absorbing agent is limited because of their low melting-point, such as DPB and DPPE. The study found that DEB' melting-point was 179℃ and its hydrogen storage capacity was good.

References

1. R. L. Courtney, L. A. Harrah, Organic Hydrogen Getters Part I: Introductory Report, SLA(1974)
2. R. D. Anderson, R. L. Courtney, L. A. Harrah, Low temperature, low pressure hydrogen gettering. U.S. Patent 3896042
3. R. M. Balooch, W. E. Wang, J. D. LeMay, Thermochemical properties of the hydrogen getter DEB, Journal of Nuclear Materials, **270**, 248-252 (1999)

4. M. Stone, M. Benson, C. Orme, et al. *Improved Hydrogen Gas Getters for TRU Waste*, INEEL/EXT-02-00564, (2002)
5. G. Mroz, *Test plan for hydrogen getters project-phase II*, LA-UR-99-1219, (1999)
6. C. A. Colmenares, R. G. Garza, *Equilibrium vapor pressures of the hydrogen getters DPB and DPPE*. UCRL-85972, (1981)
7. G. L. Powell, Hydriding kinetics of an organic hydrogen getter-DPB, Journal of Alloys and Compounds, 446–447: 402–404(2007)
8. M. Balooch, W. E. Wang, J. R. Kirkpatrick, Hydrogen uptake mechanism of a silicone-rubber DEB getter mixture, Journal of Polymer Science: Part B: Polymer Physics, **39**, 425-431 (2001)
9. Hai Zhenwei, Tang Jiansheng, Gao Zhenheng, Studies on 1,4-Bis (Phenyl Ethynyl) Benzene and Its Bis Substituted Phenyl Derivatives, chemical journal of Chinese universities, **11**(6): 586-590 (1990)
10. P. Nguyen, Synthesis of symmetric 1,4-bis(p-R-phenylethynyl)benzenes via palladium/copper catalyzed cross coupling and comments on the coupling and comments on the coupling of aryl halides with terminal alkynes, Inorganica Chimica Acta, **220**, 289-296(1994)
11. K. VSalazar, D. W. Carroll, M. Trkula, et al. Hydrogen uptake on film surfaces produced by a unique codeposition process, Applied Surface Science , **214**, 20-26(2003)
12. D. W. Carroll, D. R. Pesiri, K. VSalazar, et al. Codeposition of the DEB-Palladium hydrogen getter system, LA-UR-00-3067, (2000)
13. J. P. Kaszuba, E. Peterson,E. Mroz, et al. *Development of Hydrogen Gas Getters for TRU Waste*, WM-4237, WM'04 Conference, (2004)
14. R. L. Holtz, V. Provenzano, M. A. Imam. Overview of Nanophase Metals and Alloys for Gas Sensors, Getters, and Hydrogen Storage, Nanostructured Materials, **7**, 259-264(1996)
15. L. Rodrigo, J. A. Sawicki, R. E. Johnson, Characterization of Deactivated Metal Getters Used in a Glove Box Purification System, Fusion Science & Technology, **28**, 1410-1415(1995)
16. T. A. Giorgi, B. Ferrario, B. Storey, An Updated Review of Getters and Bettering, J. Vat. Sci. Technol., **3**, 417-423(1985)
17. F. A. Cotton, G. Wilkinson, *Transition-Metal Hydrides, Advanced Inorganic Chemistry*, 3rd Edition, Interscience Publishers, New York, 186-187(1972)
18. J. O. Henrie, D. J. Flesher, G. J. Quinn, et al. Hydrogen Control in the Handling, Shipping, and Storage of Wet Radioactive Waste, GEND-052, U.S. Department of Energy, (1986)
19. C. Carpetis. Hydrogen-Ttitium Getters and Their Applications, KCP-613-4410, Allied-Signal Aerospace Co., (1990)
20. F. A. Lewis, *The Palladium-Hydrogen System*, Academic Press, (1967)
21. L. K. Heung, Tritium Transport Vessel Using Depleted Uranium, Fusion Tech., **28**, 1385-1390(1995)
22. U. Tamm, E. Hutter, G. Neffe, et al. Uranium Getters for Tritium Cleanup at the Tritium Laboratory Karlsruhe (TLK), Fusion Tech., **21**, 983-987(1992)
23. M. Devillers, M. Sirch, S. Bredendiek-Kamper, et al. Characterization of the ZrCo-Hydrogen System in View of Its Use for Tritium Storage, Chem. Mater., **2**, 255-262(1990)

24. T. Woyke, C. Schiller, U. Schmidt, et al. ZrCo as a New H_2 Storage and Getter for Lyman-α Radiation Sources, Appl. Optics, **34**, 155-158(1995)
25. R. D. Penzhorn, M. Devillers, M. Sirch, Evaluation of ZrCo and Other Getters for Tritium Handling and Storage, J Nucl. Mater., **170**, 217-231(1990)
26. G. Bonizzoni, A. Conte, G. Gatto, et al. Tritium Storage Plant Based on a Combination of St707 and St737 Getter Alloy Beds for High Field Fusion Machines, Vacuum, **41**, 1500-1502(1990)
27. F. Doni, C. Boffito, F. Farrario, Hydrogen Isotope Sorption and Recovery by a Nonevaporable Getter Combined with a Chemical Compressor Material, J Vat. Sci. Tech., **4**, 2447-2451(1986)
28. A. N. Perevezentsev, A. C. Bell, R. Lalsser, et al. Safety Aspects of Tritium Storage in Metal Hydride Form, Fusion Technology, **28**, 1404-1409(1995)
29. S. Srivatava, O. N. Srivastava, Synthesis, Characterization and Hydrogenation Behavior of Composite Hydrogen Storage Alloys, LaNi5/La2Ni7, LaNi3, Journal of Alloys & Compounds, **281**, 197-205(1999)
30. A. Chambers, C. Park, R. T. K. Baker, et al. Hydrogen Storage in Graphite Nanofibers[J], J. Phys. Chem. B, **102**, 4253-4257(1998)
31. F. Diederich, F. L. Whetten, C60: From Soot to Superconductor, Angew. Chem. Int. Ed., **30**, 678-683(1991)
32. M. Sane, H. Nishimura, K. Ichimura, Dissociation of Hydrogen Molecules by a First-Stage Graphite-Caesium Intercalation Compound, Synthetic Metals, **31**, 73-78(1989)
33. K. Ichimura, M. Sano, Gettering of Hydrogen and Oxygen by Alkali-Metal Graphite Intercalation Compounds, Journal of Vacuum Science & Technology A Vacuum Surfaces & Films, **10**, 543-546(1992)
34. K. Janberg, F. Petrucci. Dry Storage in Casks at the Site of Super-Phenex- The Special Problem of the Tritium Getter-Process Within a Transport and Storage Cask Filled With Absorber Rods[J], in the Proceedings of ICEM'95, **1**, 285-287(1995)

Determination of Nickel in End-Fitting of Spoolable Reinforced Plastic Line Pipe

Xiaodong Shao[1,2] *

[1]State Key Laboratory for Performance and Structure Safety of Petroleum Tubular Goods and Equipment Materials, No. 89 Jinyeer Road, Xi'an 710077, China
[2]CNPC Tubular Goods Research Institute, No. 89 Jinyeer Road, Xi'an 710077, China

Abstract. Spoolable reinforced plastic line pipe will be widely used in the exploitation of acid gas and oil wells. End-fitting is a mechanical device that forms the transition from the spoolable reinforced plastic line pipe to the connector. It was well known that nickel was an important element in the end-fitting. In this paper, a novel spectrophotometric method was described for determination of nickel in the end-fitting of spoolable reinforced plastic line pipe. The method was based on the color reaction between nickel and dimethylglyoxime, which was favored in an alkaline medium, producing a red compound in the presence of ammonium persulfate as an oxidant. Under the optimum reaction conditions the absorption value was proportional to the concentration of nickel in the range of 0.080~0.50% (R2 = 0.9999), and the relative standard deviation was less than 3.0% (n=5). The proposed method was applied successfully to determine nickel in end-fitting of spoolable reinforced plastic line pipe.

1 Introduction

The condition of gas and oil wells in the west of China has complex characteristics such as deep, high pressure, and serious corrosive that only corrosion-resistant pipes can be applied to explore. One key feature there is obvious existence of corrosive hydrogen sulfide medium. In order to meet the exploitation requirements of these oil-gas wells with high pressure and high partial pressure of H_2S and CO_2, the pipes must adopt new materials with excellent resistance to corrosion. Spoolable reinforced plastic line pipe (**Fig. 1**), also known as

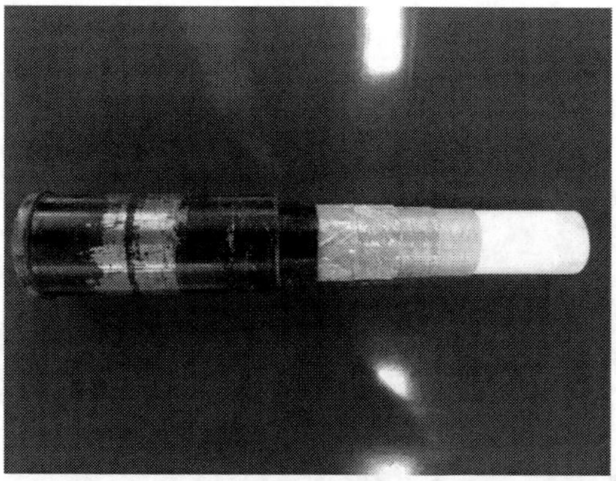

Fig. 1. Spoolable reinforced plastic line pipe

* Corresponding author: shaoxd@163.com

flexible composite pipes or reinforced thermoplastic pipes (RTP), contain three layers, among which, neck bush and reinforcement layer are keys to determine mechanical and thermal properties of these pipes. These composite pipes are key to enabling the industry to meet some of the energy security challenges that nations face today. The ability to combine tailored strength with flexibility, low weight, high thermal and pressure tolerance, minimal flow resistance, high resilience and durability in large-bore pipes enables them to surpass the capabilities of current predominantly metal piping-albeit often polymer lined. The spoolable reinforced plastic line pipe has excellent resistance to corrosion. For example, glass fibre reinforced polyoxymethylene and polypropylene (PP) performed well in saltwater, while carbon fibre reinforced polyetheretherketone proved well suited to a sour oil environment [1]. The fact that spoolable reinforced plastic line pipe is more flexible and spoolable than steel, and is also much lighter, means that it can be carried and deployed from smaller vessels and managed more easily. End-fitting is a mechanical device that forms the transition from the spoolable reinforced plastic line pipe to the connector. In order to control the quality of end-fitting products, the contents of trace impurities and alloy elements in end-fitting should be determined. It is well known that nickel is an important element in end-fitting. Therefore, a rapid and precise analytical method for the determination of nickel is essentially required for the production control of spoolable reinforced plastic line pipe. Numerous analytical methods were employed for the determination of nickel in different types of matrices including atomic absorption spectrometry [2,3], atomic emission spectrophotometry [4], molecular fluorescence spectroscopy [5], electroanalytical method [6] and spectrophotometry [7,8]. Spectrophotometric methods occupy special position due to their simplicity, less expensive instrumentation and high sensitivity. Similarly, spectrophotometric methods also play an important role in nickel determination. This present method was based on the color reaction between nickel and dimethylglyoxime, which was favored in an alkaline medium, producing a red compound in the presence of ammonium persulfate as an oxidant. The characteristic wavelength of maximum absorption was measured at 530 nm. Under the optimum reaction conditions the absorption value (A) was proportional to the concentration of nickel in the range of 0.080% ~ 0.50% (R^2 = 0.9999, mass fraction), and the relative standard deviation (RSD) was less than 3.0% (n=5). The proposed method was free from the interference from a large number of analytical important elements and has been applied satisfactorily to the determination of nickel in end-fitting of spoolable reinforced plastic line pipe samples with the improved accuracy and precision, respectively. Results were in agreement with those attained by inductively coupled plasma atomic emission spectrophotometry at the 95% confidence level. The satisfactory performance in the determination of nickel in end-fitting of spoolable reinforced plastic line pipe demonstrated that the method was practical and suitable not only for quality control analysis but also for product analysis, confirming the promise for spoolable reinforced plastic line pipe research.

2 Experiment

2.1 Apparatus and reagents

A UV-visible spectrophotometer (Model TU-1901, Beijing Persee General Instrument Co. Ltd.) with 2.0-cm quartz cells was employed for all absorbance measurements.

All chemicals including perchloric acid (Tianjin Tianda Reagent Plant), nitric acid (Xi'an Chemical Reagent Plant), ethanol (Tianjin Dongfang Reagent Plant) and hydrochloric acid (Sichuan Xilong Chemical Co. Ltd.) used in the experiments were of analytical reagent grade and were used as received. Solutions of ammonium persulfate

(Tianjin Tianda Reagent Plant), natrium tartrate (Xi'an Chemical Reagent Plant) and sodium hydroxide (Institute of Shanghai Fine Chemical Materials) were prepared daily by dissolving the reagents in deionised water that treated with a Milli-Q water purification system (Millipore, Bedford, MA, USA). Solution of dimethylglyoxime (Xi'an Chemical Reagent Plant) were prepared dissolving the reagents in ethanol, and the solutions of nickel were prepared from the steel certified reference materials (CRM) including GBW 01301, GBW 01302, GBW 01304, GBW 01306, GBW 454 (Institute of Anshan Iron and Steel Group Corporation), GBW 01354 (Institute of Fushun Iron and Steel Group Corporation). All working strength solutions in the experiment were prepared with deionised water.

2.2 General procedures

Weigh accurately 0.2000 g of drilled low alloy steel samples and dissolve the samples in 10 mL mixture acid solution of hydrochloric acid 25% (V/V) and nitric acid 25% (V/V) in a 150-mL taper bottle. Continue the treatment with mixture acid till the sample goes into solution. Add 5.0 mL perchloric acid (ρ 1.67 $g \cdot mL^{-1}$) and heat to fumes of perchloric acid, and evaporate almost to dryness. Cool the solution, add 50 mL deionised water to dissolve some salts, and cool the solution at room temperature. Finally, dilute the solution in a 100-mL volumetric flask, followed make up to the mark with deionised water, mix well and let stand. Then, transfer the above stock solution 10.0 mL in a 50-mL volumetric flask, add 10 mL of 300.0 $mg \cdot mL^{-1}$ solution of natrium tartrate, 10.0 mL of 100.0 $mg \cdot mL^{-1}$ solution of sodium hydroxide, 2.0 mL of 10.0 $mg \cdot mL^{-1}$ solution of dimethylglyoxime、 5 mL of 40.0 $mg \cdot mL^{-1}$ solution of ammonium persulfate, followed make up to the mark with deionised water, mix well and let stand. Then, transfer the above stock solution in 2.0-cm quartz cells, this solution was the work solution. At the same time, transfer the first stock solution 10.0 mL in a 50-mL volumetric flask, add 10 mL of 300.0 $mg \cdot mL^{-1}$ solution of natrium tartrate, 10.0 mL of 100.0 $mg \cdot mL^{-1}$ solution of sodium hydroxide, 2.0 mL of 95% (V/V) ethanol、 5 mL of 40.0 $mg \cdot mL^{-1}$ solution of ammonium persulfate, followed make up to the mark with deionised water, mix well and let stand. Then, transfer the above stock solution in 2.0-cm quartz cells, this solution was employed as the blank solution in subsequent experiments. Measure the absorbance at 530 nm using a UV-visible spectrophotometer with 2.0-cm quartz cells against a reagent blank prepared similarly. Plot the amount of nickel in the sample solution against absorbance to obtain the calibration graph.

3 Results and discussion

3.1 Spectral characteristics

The absorption spectra of the reagents blank, nickel against the reagent blank were tested from 400 nm to 700 nm. The nickel yields the characteristic wavelength of maximum absorption at 530 nm wavelength. However, the absorption spectrum of the reagents blank under similar conditions shown that it does not absorb at 530 nm. And the experimentation also demonstrated that nickel increases the absorbance value considerably, resulting in increased sensitivity of the method. Therefore, measure the absorbance of nickel at 530 nm.

3.2 Select dissolved acid for low alloy steel samples

The effect of the different acids for dissolve the low alloy steel samples was examined. It was found that a solution mixed with 25% (V/V) hydrochloric acid and 25% (V/V) nitric acid has a good effect for dissolve the low alloy steel samples. Thus, an acid solution mixed with 25% (V/V) hydrochloric acid and 25% (V/V) nitric acid was recommended for dissolve the low alloy steel samples in subsequent studies.

3.3 Influence of the amounts of chromogenic reagent

Dimethylglyoxime was selected as the chromogenic reagent in the proposed method. The effect of the chromogenic reagent was examined by measuring the absorbance of solution containing certain amounts of nickel and variable amounts of dimethylglyoxime. It was found that 2.0 ml of 10.0 $mg \cdot mL^{-1}$ dimethylglyoxime solution sufficed to complex the amounts of nickel taken; with higher concentrations the absorbance was essentially constant. Two milliliters of 10.0 $mg \cdot mL^{-1}$ dimethylglyoxime solution were recommended as a suitable amount of chromogenic reagent.

3.4 Influence of the amounts of oxidant

Ammonium persulfate was selected as the oxidant in the proposed method. The effect of the oxidant was examined by measuring the absorbance of solution containing certain amounts of nickel and variable amounts of ammonium persulfate. It was found that 5.0 mL of 40.0 $mg \cdot mL^{-1}$ solution of ammonium persulfate sufficed to oxidate the amounts of nickel taken, with higher concentrations the absorbance was essentially constant. Five milliliters of 40.0 $mg \cdot mL^{-1}$ ammonium persulfate solution were recommended as a suitable amount of oxidant.

3.5 Stoichiometry and stability constant

The stoichiometry of the nickel complex with dimethylglyoxime in solution was studied spectrophotometrically by Job's method of continuous variation and equilibrium shift method. The results indicate that the molar ratio of nickel to dimethylglyoxime is 1: 4. The molar extinction coefficient (ε) of the ligand complex is 1.3×10^4 $L \cdot mol^{-1} \cdot cm^{-1}$, which calculated from these results following the described method.

3.6 Interference studies

The extent of interference by diverse ions was determined by measuring the absorbance of solutions containing nickel and various amounts of diverse ions. The criterion for interference was an absorbance value varying by more than +2% from the expected value of nickel alone. The results show that a large excess of cations and anions which are usually associated in the determination of nickel do not interfere. The colorless metal ions do not interfere. The tolerable amounts of foreign species with respect to nickel for interference at +2% level were over 200~300 for the colored metal ions. When the colored metal ions contents of not more than 200~300 times nickel, it can be in some color solution adding a few drops of EDTA solution or sodium nitrite solution to fade the color, this solution as the measured reference solution to eliminate the interference. There was not obvious interference ions exist in end-fitting of spoolable reinforced plastic line pipe.

3.7 Preparation of the calibration graph

Under the optimal conditions, a series of standard solutions of nickel were prepared by the different steel certified reference materials in a 100-mL volumetric flask. add 10 mL of 300.0 $mg \cdot mL^{-1}$ solution of natrium tartrate, 10.0 mL of 100.0 $mg \cdot mL^{-1}$ solution of sodium hydroxide, 2.0 mL of 10.0 mg·mL^{-1} solution of dimethylglyoxime、5 mL of 40.0 mg·mL^{-1} solution of ammonium persulfate, followed make up to the mark with deionised water, mix well and let stand. Then, transfer the above stock solution in 2.0-cm quartz cells, this solution was the work solution. At the same time, transfer the first stock solution 10.0 mL in a 50-mL volumetric flask, add 10 mL of 300.0 mg·mL^{-1} solution of natrium tartrate, 10.0 mL of 100.0 mg·mL^{-1} solution of sodium hydroxide, 2.0 mL of 95% (V/V) ethanol、5 mL of 40.0 mg·mL^{-1} solution of ammonium persulfate, followed make up to the mark with deionised water, mix well and let stand. Measure the absorbance at 530 nm using a UV-visible spectrophotometer with 2.0-cm quartz cells against a reagent blank. The absorbance of the solution was proportional to the concentration of nickel in the range of 0.080%~0.50% (mass fraction). Plot the amount of nickel in the sample solution against absorbance to obtain the calibration graph. The linear regression equation for nickel was $A = 1.20752C + 0.00198$, $R^2 = 0.9999$, and the relative standard deviation (RSD) was less than 3.0% (n=5).

4 Applications

4.1 Determination of nickel in end-fitting of spoolable reinforced plastic line pipe samples

The recommended procedure has been applied satisfactorily to the determination of nickel in end-fitting of spoolable reinforced plastic line pipe samples. The samples were dissolved according to the following procedure. Measure the absorbance at 530 nm using a UV-visible spectrophotometer with 2.0-cm quartz cells against a reagent blank. The results were summarized in Table 1 with the good accuracy and precision.

Table 1. Results of determination of nickel in end-fitting samples (ω, n=5)

Sample No.	*A*	RSD (%)	Content (%)	
			Proposed method	AES
1	0.4488	0.68	0.37	0.37
2	0.4608	1.56	0.38	0.39
3	0.4488	2.03	0.37	0.37
4	0.4729	1.21	0.39	0.38
5	0.4729	1.88	0.39	0.40
6	0.4367	0.93	0.36	0.36
7	0.1469	2.54	0.12	0.11
8	0.1227	1.47	0.10	0.11
9	0.1348	2.33	0.11	0.12
10	0.1590	1.64	0.13	0.13
11	0.1227	1.35	0.10	0.10
12	0.1227	1.96	0.10	0.11
13	0.2435	1.01	0.20	0.19
14	0.2556	0.98	0.21	0.20
15	0.2556	1.24	0.21	0.21
16	0.2918	2.18	0.24	0.25
17	0.2676	1.57	0.22	0.22
18	0.2556	1.39	0.21	0.20

5 Conclusions

Based on the chromogenic reaction between nickel and dimethylglyoxime, a simple and accurate spectrophotometric method for determination of nickel in end-fitting of spoolable reinforced plastic line pipe was proposed. The red colored nickel solution exhibits absorption maximum at 530 nm. The proposed method was free from the interference from a large number of analytical important elements and has been applied satisfactorily to the determination of nickel in end-fitting of spoolable reinforced plastic line pipe samples. Compared with other methods for the determination of nickel, this method offers advantages of simplicity, less expensive instrumentation and high efficient. The satisfactory performance in the determination of nickel in end-fitting of spoolable reinforced plastic line pipe demonstrated that the method was practical and suitable not only for quality control analysis but also for product analysis, confirming the promise for spoolable reinforced plastic line pipe research.

The authors gratefully acknowledge the State Key Laboratory for Performance and Structure Safety of Petroleum Tubular Goods and Equipment Materials, and CNPC Tubular Goods Research Institute.

References

1. K. Yu, E.V. Morozov, M.A. Ashraf, K. Shankar, Composite Structures **131**, 453 (2015)
2. M. Pozzatti, B. Welz, Microchemical Journal **133**, 162 (2017)
3. B.M. Soares, R.F. Santos, F.A. Duarte, Talanta **160**, 454 (2016)
4. L. Li, H. Gao, J. Zhang, Metallurgical Analysis **27**, 41 (2007)
5. M. C. Talio, M. Alesso, M. Acosta, V. S. Wills, L. P. Fernández, Talanta **174**, 221 (2017)
6. E. Punrat, P. Tutiyasarn, O. Chailapakul, Talanta **168**, 286 (2017)
7. B.N. Kumar, S. Kanchi, M.I. Sabela, K. Bisetty, Karbala International Journal of Modern Science **2**, 239 (2016)
8. Y. Guo, H. Zhao, Q. Zhang, X. Bian, Spectrochimica Acta Part A **173**, 532 (2017)

Cyanide Containing Wastewater Treatment by Ozone Enhanced Catalytic Oxidation over Diatomite Catalysts

Mingguo Lin[1], *Qiyuan* Gu[2], *Xinglan* Cui[2] and *Xingyu* Liu[2 *]

[1]Liaoning Xindu Gold Co., Ltd, 122000 Chaoyang, China
[2]General Research Institute for Nonferrous Metals Engineering, National Engineering Laboratory of Biohydrometallury, 100088 Beijing, China
The first two authors contributed equally to this paper

Abstract. Cyanide containing wastewater that discharged from gold mining process creates environmental problems due to the toxicity of cyanide. As one of the promising advanced oxidation process, catalytic oxidation with ozone is considered to be effective on the purification of cyanide. Diatomite, a natural mineral, was used as catalyst in this study. The effect of O_3 dosage, salinity, initial cyanide concentration and initial pH condition were investigated. It was observed that the removal rate of cyanide was much higher in the catalytic oxidation with ozone process than the one in zone alone process. Alkaline condition was especially favorable for cyanide in catalytic oxidation with ozone. The ozone and catalytic oxidation with ozone were simulated by pseudo-first-order kinetics model. The apparent first-order rate constant contribution of the diatomite catalyst was 0.0757 min^{-1}, and the contribution percentage was 65.77%.

1 Introduction

Cyanidation is the most popular process by which gold is recovered from the ores in mining operations. The aqueous cyanide solution which is used to dissolution of gold is typically carried out with 0.03-0.3% NaCN [1]. The cyanide in the solution is used as complexing agent. However, after extraction and recovery of gold, the drainage containing large amounts of cyanide need be discharged, which create environmental problems due to the toxicity of cyanides [2].

The cyanide could present three forms in wastewater, including free cyanide, simplex cyanide and complex cyanide. For the widespread use of cyanide in gold mining operation, the purification of cyanide is important. Both treatment efficiency and cost could be considered [3]. The Ministry of Environmental Protection of the People's Republic of China has proposed a discharge limit of 0.5 mg/L cyanide for sewers. And in Liaoning Province, the limit for cyanide is 0.2 mg/L. For meeting the strict discharge standard requirement, several cyanide treatment processes have been developed.

* Corresponding author: wellwoodliu@163.com

Catalytic oxidation with ozone is one of the promising treatment techniques for cyanide removal. Ozone has been demonstrated to be effective on the purification of non-biodegradable compounds, by the way of direct oxidation with O_3 and the indirect way with hydroxyl radicals [4].

In the last years, the combined use of O_3 and activated carbon has been developed [5]. The treatment efficiency was enhanced dramatically by using activated carbon as catalyst [6]. Unfortunately, the cost of activated carbon is expensive, which is limited the application of the combined process. Diatomite could be the substitute catalyst which has porous structure and high surface area. In this study, the operating parameter of combined O_3 and diatomite were optimized.

2 Materials and methods

2.1 Wastewater

The real cyanide containing wastewater was captured from a gold mine company in Liaoning Province. The cyanide concentration is 15~20 mg/L, and COD concentration is 150~200 mg/L.

2.2 Analyses

The concentrations of cyanide were measured by using an ion selection electrode. COD was measured according to standard methods. The microstructure of diatomite was observed by scanning electron mircroscope (ZEISS EVO 18). The increase in apparent rate constant (pseudo-first-order) K_{app} resulted from adding diatomite during the ozone process can be used to indicate the contribution and contribution percentage of effect of diatomite catalyst, which was calculated as the following equation:

$$Contribution\ K_{app} = K_{app}(with\ diatomite) - K_{app}(without\ diatomite) \quad (1)$$

$$Contribution\ percentage = (1 - K_{app}(without\ diatomite)/K_{app}(with\ diatomite) \times 100\% \quad (2)$$

The pseudo-first-order kinetics of both processes was simulated as:

$$In(\ C_0/C) = K_{app}t \quad (3)$$

Where C is the cyanide concentration at time t, C_0 is the initial cyanide concentration, and K_{app} is the apparent first-order rate constant.

2.3 Experimental method

The schematic diagram of combined ozone and diatomite system is shown in Fig.1. Two parallel set of ozone reactor, named R1 and R2, were operated concurrently. R2 was similar with R1 except adding diatomite. R1 and R2 reactor were made by borosilicate glass. Ozone was continuously added to the reactor.

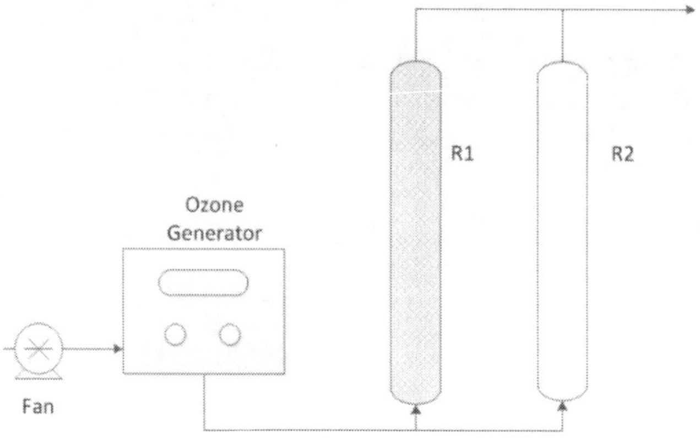

Fig. 1. Schematic diagram of combined ozone and diatomite system.

3 Results and discussion

3.1 Micro-morphology observation of diatomite

The electronic micrograph of diatomite was shown in Fig. 2. The diatomite has high porosity which could provide the site to promote the reaction between O_3 and cyanide.

Fig. 2. Scanning electron micrographs for diatomite.

3.2 Effect of O_3 dosage

Fig.3 shows the cyanide concentration as a function of O_3 dosage. The results indicated the combined O_3 and diatomite process presented better cyanide purification performance than the O_3 alone process. Cyanide concentration in R1 decreased from 15.8 mg/L to 0.1 mg/L with an increase in the O_3 dosage to 16 mg/L, while the cyanide concentration in R2 dropped to 6.3 mg/L.

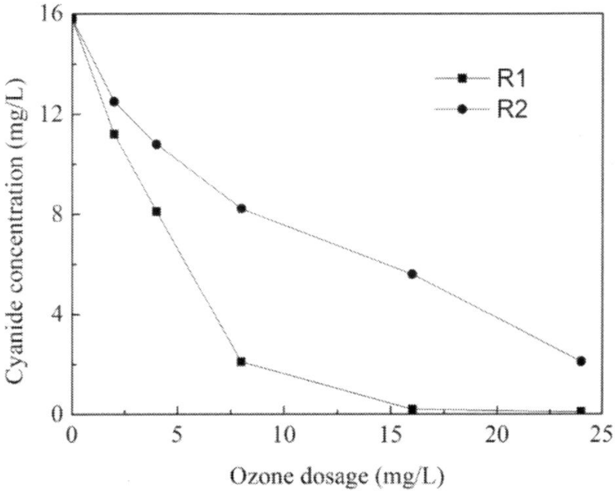

Fig. 3. Effect of O_3 dosage on the removal of cyanide.

3.3 Effect of salinity

Salinity always influences the catalytic oxidation via the competitive adsorption of inorganic anions on catalyst sites. The influence of NaCl was investigated during catalytic oxidation with ozone. The effect of NaCl is shown in Fig. 4, in which a series amount of NaCl (0, 0.5 g/L, 1 g/L) was added in the oxidation system. It is observed that the salinity did not influence the catalytic oxidation significantly. The real industrial wastewater always contains high salinity. The results indicated that the catalytic oxidation with ozone process could be suitable for the treatment of industrial wastewater.

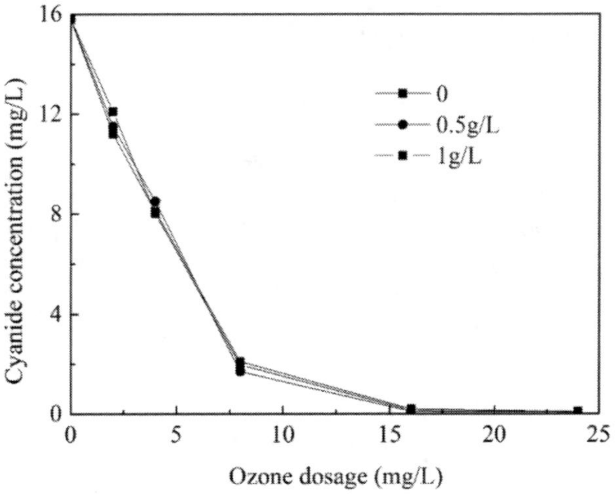

Fig. 4. Effect of salinity on the removal of cyanide.

3.4 Effect of initial pH

Cyanide containing wastewater could influence the charge property of cyanide and removal efficiency of ozone process. The effect of cyanide containing wastewater pH on the treating rate was investigated from 5.0 to 9.0. As illustrated in Fig.5, the fastest cyanide removal rate was observed in the condition of pH 9.0, while the slowest was demonstrated in the condition of pH 5.0. The acid solution condition could not promote the reaction activity between O3 and cyanide. These observation also indicated that at high pH the hydroxyl radical could be the predominant oxidant for the cyanide. The product of cyanide oxidation is cyanate, which occurs according to Eq(4):

$$CN^- + O_3 \rightarrow CNO^- + O_2 \tag{4}$$

In the mostly alkaline solutions hydrolysis of CNO^- is also possible:

$$CNO^- + OH^- + H_2O \rightarrow CO_3^{2-} + NH_3 \tag{5}$$

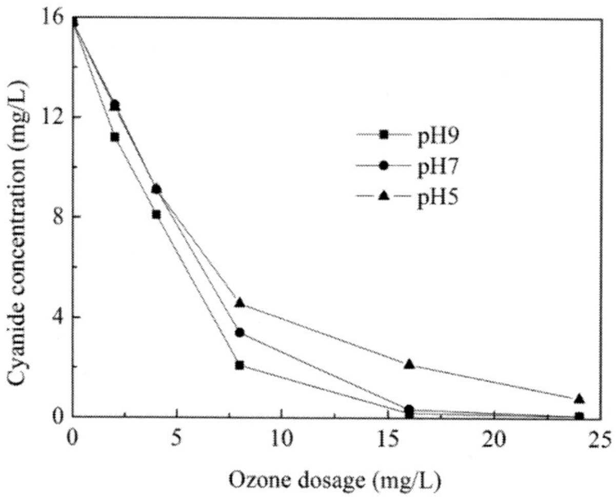

Fig. 5. Effect of pH on the removal of cyanide.

3.5 Effect of initial cyanide concentration

The tests of difference initial cyanide concentrations were carried out (Fig. 6). When the initial cyanide concentration was 15.8 mg/L, 16 mg/L O3 was needed to add into the solution. The initial cyanide removal rate decreased with the drop of the initial cyanide concentration. High initial cyanide concentration could promote the oxidation reaction. The increased initial cyanide concentration provided more drive force to transfer cyanide to diatomite surface. Therefore, the hydroxyl radicals generated by ozone could react with more cyanide.

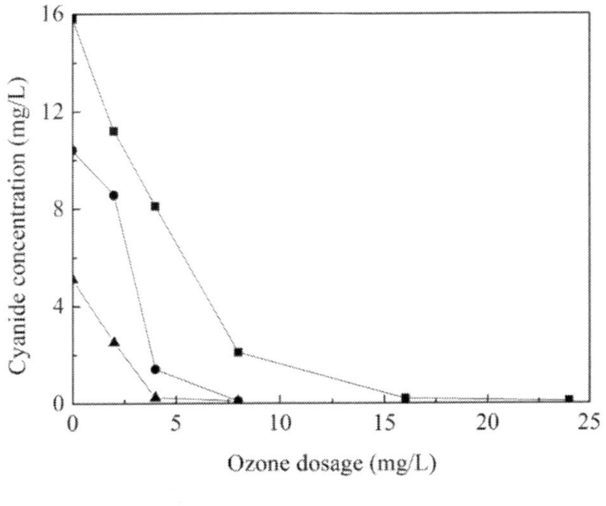

Fig. 6. Effect of initial concentration on the removal of cyanide.

3.6 Kinetics analysis

The pseudo-first-order kinetics was used to simulate the cyanide degradation data. Fig.7(a) shows the change trend of cyanide removal efficiency vs time. The apparent first-order rate constants of cyanide oxidation were determined from the slope of the straight lines obtained by plotting the ratios of the remaining concentration to the initial concentration on logarithmic scale. As shown in the Fig. 7(b), the data could fit the pseudo-first-order kinetics model very well. It is also evident from Fig. 7(b), that the behavior of cyanide oxidation identified a pseudo-first-order reaction with respect to cyanide concentration in solution. The apparent first-order rate constant of the ozone oxidation without diatomite is 0.0394 min-1, and regression R2 is 98.41%, while the one with diatomite is 0.1151 and 98.07%. The cyanide oxidation was enhanced by adding diatomite as catalyst. The catalyst of diatomite during the ozone oxidation plays a significant role during the process. The K_{app} contribution in the present of diatomite was 0.0757 min^{-1}, and the contribution percentage of adding diatomite was 65.77%.

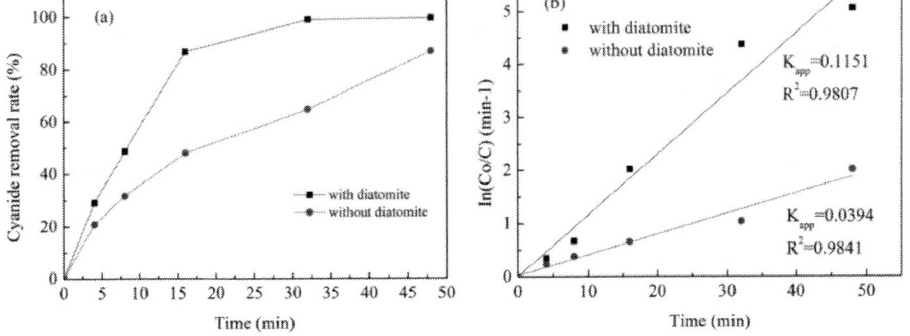

Fig. 7. Pseudo-first-order kinetics modeling: (a) trend of cyanide removal efficiency vs time, (b) correlation between experimental data (points) and prediction curves (line).

4 Summary

During cyanide oxidation by catalytic oxidation with ozone, the contribution and contribution percentage of diatomite which was used as catalyst were investigated by comparison with single ozone process. The apparent first-order rate constant contribution of the diatomite catalyst was 0.0757 min^{-1}, and the contribution percentage was 65.77%. In addition, the effect study of O_3 dosage, salinity, initial cyanide concentration and initial pH condition were carried out. It was observed that alkaline condition was especially favorable for cyanide in catalytic oxidation with ozone. And the solution salinity did not influence the catalytic oxidation effect. These results indicated that catalytic oxidation with ozone is effective on the purification of cyanide.

This work was supported by the fund of General Research Institute for Nonferrous metals under grant numbers 53321 and 53348, and the public welfare fund of the Ministry of Environmental Protection of People's Republic of China under grant number 201509049, the National Natural Science Foundation of China under grant numbers U1402234, 41573074.

References

1. J. A. Zeevalkink, D. C. Visser, P. Arnoldy, C. Boelhouwer, Wat. Res. **14,** 1375 (1980)
2. J. R. Parga, D. L. Cocke, Desalination **140,** 289 (2001)
3. U. Kepa, E. S. Mazanek, L. Stepniak, Desalination, **223,** 187 (2008)
4. M. D. Gurol, R. Vatistas, Wat. Res. **21,** 895 (1980)
5. F. R. Carrillo-Pedroza, F. Nava-Alonso, A. Uribe-Salas, Miner. Eng. **13,** 541 (2000)
6. J. Cui, X. Wang, Y. Yuan, X. Guo, X. Gu, L. Jian, Chem. Eng. J. **241,** 184 (2014)

Two Water Stable Copper Metal-Organic Frameworks with Performance in the Electrocatalytic Activity for Water Oxidation

Xiuping Liu[1] , *Yijun* Wang[2] , *Wenwen* Liu[1] , *Lianming* Zhao[1 *], and *Wenyue* Guo[1,*]

[1]College of Science, China University of Petroleum, Qingdao, Shandong 266580, People's Republic of China

[2]Linyi jun cheng leather co., LTD, Linyi, Shandong 276000, People's Republic of China

Abstract. Two novel water stable metal-organic frameworks, $[Cu(\mathbf{L})\cdot(4,4'\text{-bipy})\cdot(ClO_4)]_n$ (**1**), $[Cu(\mathbf{L})\cdot(phen)\cdot(ClO_4)\cdot(H_2O)]_2$ (**2**), have been constructed by **HL**=[5-Mercapto-1-methyl] tetrazole acetic acid and Cu (II) salt in the presence of assistant N-containing ligands. MOF **1** and MOF **2** with open Cu^{II} sites, resulting the framework **1** and **2** show electrocatalytic activity for water oxidation in alkaline solution. The electrochemical properties of complex for oxygen evolution reaction (OER) were evaluated by linear sweep voltammetry (LSV) and the Tafel slopes. Complex **1** has a higher LSV activity with a lower over potential of 1.54 V and a much higher increase in current density. Meanwhile, the Tafel slope of complex **1** (122.0 mV dec^{-1}) is much lower than complex **2** (243.5 mV dec^{-1}). This phenomenon makes complex **1** a promising porous material for electrocatalytic activity.

1 Introduction

Metal−organic frameworks (MOFs), as a new and vital class of porous crystalline materials, have received considerable attention in the past decades for their potential applications in the adsorption, catalysis, biology and others.[1] The N-donor molecules and theirs derivatives have been widely used as the secondary ligands because of their rigid skeletons and the structures formed are more predictable.[2] In the present paper, the ligand contains carboxylica acid and neutral four nitrogen thiazole ring.

Recently, some reports have shown that most of MOFs are unstable and lose their structural when exposed in water. Because the relatively weak metal−oxygen bonds within the frameworks are easily attacked by water molecules and lose their structures.[2] Fortunately, the incorporation of hydrophobic functional groups within the frameworks might largely enhance the M-O bonds and thus improve the water resistance of the MOFs in an efficient manner. For example, introducing a $-CH_3$ functional group within the framework might be enhance the water/moisture stability of the frameworks.[3]

Most of the reports on MOFs were focused on structures, gas-uptake, and so on, and the electrochemical reactivity of MOF systems, especially for the oxygen evolution reaction

* Corresponding author: lmzhao@upc.edu.cn

(OER) was seldom studied.[4] Therefore developing highly efficient electrocatalysts for OER has emerged as one of the research hot spots in clean energy generation and storage.[5] It is of great importance to explore the series of materials to substitute traditional cost-effective oxygen evolution catalysts. However, the well-established catalysts for water oxidation up to now are mainly recorded as some precious metals or traditional metal materials,[6] and the MOFs used for OER are nearly unexplored.[7] In this context, we present the syntheses, crystal structures and the OER properties of the two complexes, namely $[Cu(\mathbf{L})(4,4'\text{-bipy})]_n$ (**1**), $[Cu(\mathbf{L})(phen)]_n$ (**2**), which incorporates 4,4′-bipy and phen ligands respectively.

2 Experiment

2.1 Reagent and instruments

All the reagents and solvents employed were commercially available and used as received without further purification except **HL**. X-ray powder diffractions were measured on a Panalytical X-Pert pro diffractometer with Cu-Kα radiation. The elemental analyses (C, H, and N) were determined on a CE instruments EA 1110 analyzer. TG curves were measured from 40 to 900 °C on a Mettler Toledo TGA instrument at a heating rate 10 °C/min under the N_2 atmosphere (100 mL/min).

2.2 Synthesis of [5-Mercapto-1-methyl] tetrazole acetic acid

The mixture of 1.0 g 5-Mercapto-1-methyltetrazole and the solution of 0.4 g NaOH in 10mL deionized water in a reaction flask at room temperature, the solution of 1.06 g sodium chloroacetate in 20 mL deionized water was added dropwise, then the mixture was stirred at 50 °C for 5 h. After cooled, the reacting mixture was acidified by 1mol/L hydrochloric acid to pH=3.5. In this process form a precipitation. The precipitation was filtered, and followed by ethyl rinsed, dried and finally became white powder, with a yield of 90%, which was purified by recrystallization in ethanol-water to get white crystals.

2.3 Synthesis of complex (1)

Pale-blue block-shaped crystals of **1** can be easily synthesized from $Cu(NO_3)_2$ (2 mmol), 4,4'dithiosalicylic (2 mmol) and **HL** (**HL**= [5-Mercapto-1-methyl] tetrazole acetic acid in 10 ml ethanol/water (V/V= 1:1) solvents. And then were sealed in a pressure-resistant glass bottle and slowly heated to 75 °C from room temperature. After keeping at 75 °C for 3000 min, the mixture was slowly cooled to 30 °C at a rate of 7 °C/h. Blue rectangular block crystals suitable for single-crystal X-ray diffraction were obtained. The products were washed with H_2O and dried in air, 60% yield based on Cu^{II}. Elem anal. Calcd. for $C_{14}H_{13}ClCuN_6O_6S$ (%): C, 33.86; H, 2.17; N, 16.65; S, 6.46. Found (%): C, 34.12; H, 2.18; N, 17.01; S, 6.5.

2.4 Synthesis of complex (2)

Metal organic framework **2** was synthesized in an analogous procedure to **1** except that 4,4' dithiosalicylic was replaced by phen, giving blue rectangular block X-ray-quality crystals in a 65% yield. Anal. Calcd. (%) for $C_{16}ClCuN_6O_7S$ (%): C, 36.94; H, 2.75; N, 16.16; S, 6.17. Found (%): C, 35.92; H, 2.77; N, 15.58; S, 5.70. The crystal data collection and refinement processes are summarized in Table 1.

Table 1. Crystal data and structure parameters for the complexes **1** and **2**.

Complex	1	2
Formula	$C_{14}H_{13}ClCuN_6O_6S$	$C_{16}ClCuN_6O_7S$
Formula weight	492.15	519.27
Temperature/K	293 (2)	293 (2)
Crystal system	triclinic	triclinic
Space group	I2/c	P-1
a / Å	19.524(4)	7.5420 (6)
b / Å	11.14515 (19)	10.8651 (11)
c /Å	16.3628(3)	13.3290(14)
α / (°)	90	87.056(9)
β/ (°)	94.5919(17)	76.136(8)
γ / (°)	90	72.575(8)
V / Å3	3620.19 (12)	1011.51 (18)
Z	8	2
Dc / (g·cm-3)	1.519	1.519
F(000)	1992	512.0

3 Results and discussions

3.1 Crystal structure of [Cu (L)·(4,4'-bipy)·(ClO_4)]$_n$ 1

Single-crystal X-ray diffraction analysis reveals that complex **1** possess 2D lamellar structure, crystallizing in monoclin I2/c space group. As shown in Fig. 1, The asymmetric unit in complex **1** contains one crystallographic independent Cu^{II} ion, two **L** ligand, two 4,4'-bipy ligand, and one uncoordinated ClO_4 anion. Complex **1** is a monocline laser crystal. Each Cu^{II} atom is five-coordinated by two O atoms from two ligands and three N atoms from two 4,4'-bipy and one ligand. At last, the 2D infinite chains were further stacked into a three-dimensional supramolecular framework through π-π stacking and weak interactions (Fig. 1a).

3.2 Crystal structure of [Cu (L)·(phen)·(ClO_4)·(H_2O)]$_2$ 2

X-ray single-crystal diffraction reveals that complex **2** crystallized in monoclinic P-1 space group. The asymmetric unit in complex **2** contains one crystallographic independent Cu^{II} ion, two lattice water molecular, two L ligand, two phen ligand, and one uncoordinated ClO_4 anion. Each Cu^{II} ion is five-coordinated by two O atoms from one ligands and water, three N atoms from one phen and one ligand, and the adjacent two Cu^{II} centers are bridged by two ligand to form the framework (Fig. 1e).

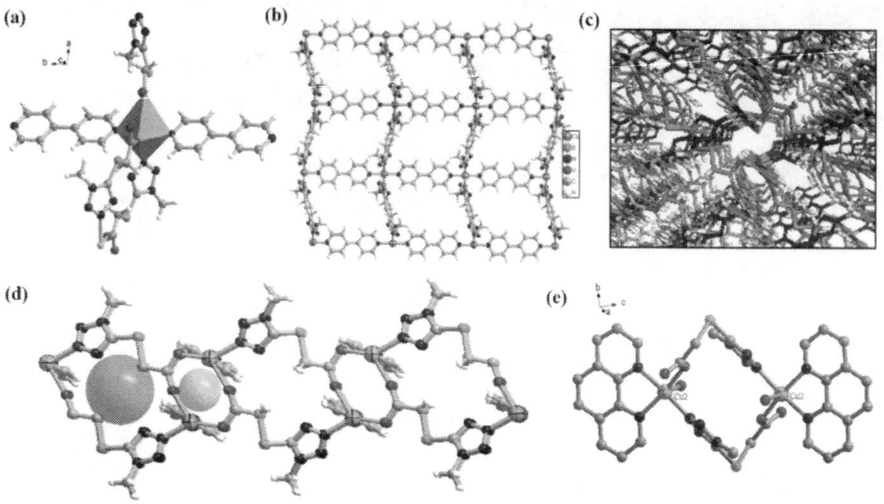

Fig. 1. (a) Coordination environment of Cu(II) center, (b) 2D sheet of **1**, (c) A perspective view along the c axis showing the hexagonal shapedchannel in **1**, (d) view of the 2D framework of **1** along the c axis showing the 1D open channels, (e) Coordination environment of Cu(II) center of **2**.

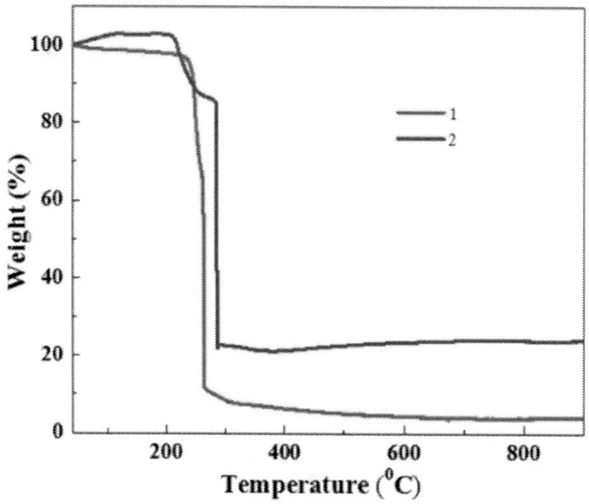

Fig. 2. TG curve of complex **1** and **2**.

3.3 Thermal analysis and powder X-ray diffraction

The thermal behavior of sample **1** and sample **2** were studied by TGA under N_2 atmosphere with a heating rate of 10 °C/min (Fig. 2). For complex **1**, up to 220 °C, **1** begins to decompose. As for complex **2**, above 270 °C, complex **1** begins to decompose. This was further confirmed by the temperature-dependent PXRD results.

Variable-temperature PXRD for **1** and **2** were measured from room temperature to 200 °C. As clearly exhibited in Fig. 3, as the temperature increases from 25 to 200 °C, the PXRD pattern remains almost unchanged in terms of both the number and positions of the peaks, indicating the unchanged framework of **1** and **2** over this temperature range. This is also true for crystals of **1** and **2** upon treatment in boiling water or soaking in alcohol. Thus the stabilities of **1** and **2** in aqueous solutions were investigated by PXRD (Fig. 3). For these tests, 40 mg of as-synthesized **1** and **2** were immersed in boiling water for an hour and soaked in aqueous solutions, methanol solutions for 2 days at room temperature. According to these results, indicating the maintain of the porous structures of complex **1** and **2** in boiling water and methanol solutions. This may be due to the hydrophobic functional groups (−CH_3) are close to the copper center within the framework of **1** and **2**, and could prevent the Cu−O and Cu−N bonds from attacking by water and thus effectively enhance the hydrophobic properties. In addition, the experimental patterns are in good agreement with simulated patterns generated from the results of single-crystal diffraction data, demonstrating the phase purity of the product nicely.

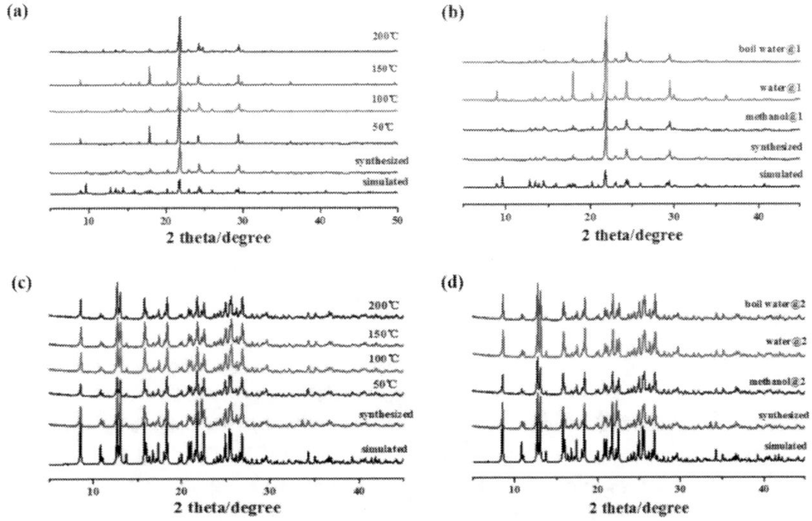

Fig. 3. PXRD patterns of **1** and (a) patterns from room temperature to 200 °C; (b) patterns of samples with distinct chemical treatments. PXRD patterns of **2** (c) patterns from room temperature to 200 °C; (d) patterns of samples with distinct chemical treatments.

3.4 Electrochemistry activity

Electrocatalytic properties for a myriad of electrochemical reactions such as OER have gained great research attention.[8] In recent years, MOFs structures were seldom reported as OER catalyst. Therefore, we study the electrocatalytic activity of the complex **1** and complex **2**. Electrochemical measurements were performed at an electrochemical workstation (Gamry Reference 600 Instruments, USA). The measurement was performed in 40 mL of 1.0 M KOH (aq) electrolyte prepared using 18 MΩ deionized water and purged with O_2 before the measurement for half an hour in order to saturate the solution. The potentials reported in this paper were converted from vs. SCE to vs. RHE according to the equation E (RHE) = E (SCE) + 1.07 V. Before measurement, the GCE was polished with 1.0, 0.3 and 0.05 μm alumina powder. Then it was ultrasonicated in deionized water and the absolute ethanol for 10 min several times. For the preparation of the sample inks, 5 mg of the as-prepared sample was dispersed in 1 mL dichloromethane solution and ultrasonicated for half an hour. Then, the slurry of 10 uL was pipetted on the surface of GCE and dried in air for use. Linear sweep voltammetry (LSV) was conducted from 1.2 V to 1.85 V vs. RHE with a scan rate of 10 mV s^{-1}. iR compensation was applied in it.

The electrocatalytic properties for OER of the two prepared samples are compared through the electrochemical measurement as is shown in Fig. 4. The linear sweep voltammetry (LSV) of the two samples are compared (Fig. 4a). It can be seen that the complex **1** has a higher LSV activity with a lower over potential of 1.54 V and a much higher increase in current density. The Tafel slopes of the two samples extracted from Fig. 1a are exhibited in Fig. 4b. The Tafel slope of complex 1 (122.0 mV dec^{-1}) is much lower than complex **2** (243.5 mV dec^{-1}), leading to an advantage for practical catalytic application. Thus, complex **1** has a much better electrocatalytic activity owing to the lower over potential and Tafel slope. This may be due to the follow reasons, first, it is known that the electrocatalytic reaction is primarily a surface phenomenon,[9] and complex **1** has a high specific surface area and offers sufficient active sites for OER, which could be verified by the program Poreblazer V.3.0.2.51 and its structure; second, the ligand is supposed to promote electron transport at the heterogeneous surface due to its excellent electrical conductive properties, and it acts as an ideal platform for loading Cu^{II}; third, as carbon- and nitrogen-based organic frameworks have showed their promising electrocatalytic properties, such as in The hydrogen evolution reaction (HER), oxygen reduction reaction (ORR), and OER.[10]

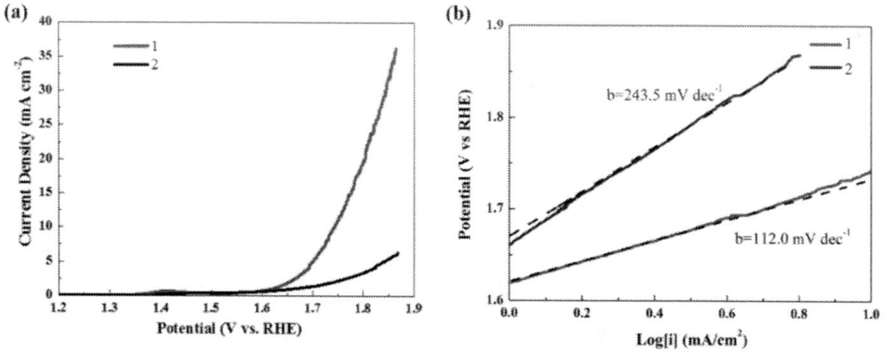

Fig. 4. Electrocatalytic activity of complex **1** and **2** for OER. (a) Polarization curves and (b) Tafel plots in 1.0 M KOH and 0.2 M PBS solution (10 mV s^{-1}).

4 Summary

In summary, two Cu-MOFs: $[Cu(\mathbf{L})\cdot(4,4\text{-bipy})\cdot(ClO_4)]_n$ (**1**), $[Cu(\mathbf{L})\cdot(phen)\cdot(ClO_4)\cdot(H_2O)]_2$ (**2**), based on ligand H**L** and $Cu(NO_3)_2$ have been successfully obtained. The incorporation of hydrophobic functional groups (methyl) on the ligands attributes to the two new water stable Cu-MOFs. Compared with those non-noblemetal electrocatalytic materials reported recently, the complex **1** electrode presents a large exchange current density and a relatively small Tafel slope. We suggesting that complex **1** is a promising low-cost and earth-abundant metallic electrocatalyst for OER.

References

1. X.P. Liu, Z.Y. Xiao, J. Xu, W.B. Xu, P.P. Sang, L.M. Zhao, H.Y. Zhu, D.F. Sun, W.Y. Guo, J. Mater. Chem. A **4**, 13844 (2016)
2. K. Tan, N. Nijem, Y. Gao, S. Zuluaga, J. Li, T. Thonhauser, Y.J. Chabal, CrystEngComm **17**, 247 (2015)
3. M.E. Braun, C.D. Steffek, J Kim, P.G. Rasmussen, O.M. Yaghi, Chem. Commun. **24**, 2532 (2001)
4. K. Sakaushi, M. Antonietti, Acc. Chem. Res. **48**, 1591 (2015)
5. C.L. Tan, H. Zhang, Chem. Soc. Rev. **44**, 2713 (2015)
6. S.S. Wang, G.Y. Yang, Chem. Rev. **115**, 4893 (2015)
7. F. Schüth, Chem. Mater. **26**, 423 (2014)

8. J.S. Li, S.L. Li, Y.J. Tang, M. Han, Z.H. Dai, J.C. Bao and Y.Q. Lan, Chem. Commun. **51**, 2710 (2015)
9. J. Xiao, Q. Kuang, S.H. Yang, F. Xiao, S. Wang and L. Guo, Sci. Rep, **3**, 2300 (2013)
10. Y. Cheng, J. Zhang and S. P. Jiang, Chem. Commun. **51**, 13764 (2015)

Preparation of Activated Carbons from Waste External Thermal-Insulating Phenolic Foam Boards

Lijuan Gao[1], *Xiaojun* Zheng[1], *Yaming* Zhu[1], *Xuefei* Zhao[1*], and *Junxia* Cheng[1]

[1] Engineering Research Center of Advanced Coal Coking and Efficient Use of Coal Resources, University of Science and Technology Liaoning, 114051 Anshan China

Abstract. Activated carbons (ACs) were prepared by steam physical activation or KOH chemical activation with the waste external thermal-insulating phenolic foam board as the raw material. The Fourier transform infrared spectroscopy (FT-IR), X-ray diffraction (XRD), Brunauer-Emmett-Teller (BET) specific area, pore-size distribution and iodine value were used to characterize the properties of ACs. AC-1(with the method of KOH chemical activation) has the iodine value of 2300mg/g, BET specific area of 1293 m^2g^{-1}, average pore-size of 2.4 nm, and mainly composed of micropore and relatively small mesopore. AC-2(with the method of steam physical activation) has the iodine value of 1665mg/g. Compared with AC-2, AC-1 had a pore-size distribution with more evenly and relative concentrated, it's belonging to the high microporosity materials. Actually, chemical activation had more significant influence on destruction of the pore wall than physical activation.

1 Introduction

As a kind of self-flame-retarded foam material, thermal-insulating phenolic foam boards are good at fire prevention and thermal insulation of buildings with its series remarkable performance. For example, hermal-insulating phenolic foam boards have the specialty of non-drop, non-soften small smoke amount, no diffusion flames, flame penetration resistance, fire retardancy and heat preservation and insulation [1, 2]. However, the recycling of waste treatment on thermal-insulating phenolic foam boards has been a new problem.

Activated carbon is an important carbon material with its good ability of high porosity, specific surface area, stable chemical property, easy regeneration and so on [3-5]. Activated carbons are extensively used as adsorbent, catalyst, catalyst carrier, gas or electric energy storage, electric double layer capacitor (EDLC) electrode material, in the fields of food, medicine, chemical industry, and environmental protection [6]. Actually, almost any carbonaceous material can be converted into activated carbon, and the preparation methods of ACs were predominately includes carbonization and activation [7-9]. Generally, the raw materials of ACs can be roughly divided into two categories of plants and minerals according

* Corresponding author: zhaoxuefei@163.com; gaolijuan@ustl.edu.cn

to the sources [10-12]. Plant category includes wood, coconut shells, walnut shells, apricot, olive nuclear, and rice husk. Minerals category includes anthracite, asphalt and petroleum coke and. In recent years, a variety of wastes have been favored to produce ACs with their outstanding advantages of low producing cost, available, sufficient, high carbon content and nonpoisonous raw materials [13-15]. It's an excellent method to save resources and protect the ecological environment produce for produce ACs from wastes.

Thermal-insulating phenolic foam board is a kind of high quality carbon materials, and its wastes can be used as raw material during the production of ACs. It's a good method to saving resources and protecting ecological environment. In this paper, the properties of ACs prepared by two activation processes were carefully compared. It takes an important role to provide basis for the fundamental research on the utility value of waste phenolic foam and ACs preparation.

2 Experiment

2.1 Pre-treatment of the raw material

A certain quantity of waste external thermal-insulating phenolic foam boards were grounded and sieved to 21 μm, then dried in 120 °C for 3 hours before use. The raw material has been named as RM. The pre-treatment waste phenolic foam was heated to 800 °C at the rate of 5 $°C \cdot min^{-1}$ with N_2 as the protcetion gas, then holding for 60 minutes, and the obtained product numbered as CP.

2.2 Preparation of activated carbon

Alkali activation. A certain amount of CP were homogeneous mixed with KOH solution (0.1 $g \cdot mL^{-1}$) at different weight ratio, and dipped for a certain hours. The rest materials were transferred to a corundum boat after eliminated the free water. Then the corundum boats were put to the activating oven for activation. Samples were naturally cooled to the ambient temperature after a certain activation time. The obtained activated samples have been cleaned by hydrochloric acid (5%) and distilled water till to neutral pH value. The obtained activated carbons were dried to constant weight and calculated yield finally.

Steam activation. The corundum boat (with 1.0000g CP) was put transferred to activated oven, and heated from ambient temperature to activation temperature at a certain rate with N_2 as the inert atmosphere. The N_2 was replaced by pre-set steam at the activation temperature. After a certain activation time, the steam was shut down and samples were cooled naturally within N_2, and then gained the activated carbon.

2.3 Characterization of samples

FTIR analysis. Infrared (IR) spectra of pre-treatment materials and ACs were obtained using a PerkinElmer Spotlight-400 FTIR/NIR spectrometer at a resolution of 4 cm^{-1}. Prior to FTIR analysis, KBr pellets were prepared by grinding the mixture of 1 mg of sample with 150 mg of KBr.

XRD analysis. The samples (RM, CP, AC-1, and AC-2) were characterized on powder X-ray Diffraction (XRD) with the tube voltage of 40 kV, tube current of 40 mA, scanning range from 10°~90°and step size of 0.013°.

BET analysis. The Brunauer-Emmett-Teller (BET) has been used to analysis the pore structure of ACs. Actually, the ACs samples were characterized at fully automatic physical adsorption instrument (ASAP2020, American). The samples have been dried for 4h at an vacuum drying oven before test.

Conventional analysis. The iodine adsorption value of activated carbon was tested using the standards of GB/T 7702.7-2008 (China).

3 Results and discussion

3.1 Effects of steam activation parameters on the adsorptive property of activated carbon

The effects of steam flow, activation temperature, and activation time on the yield and adsorptive property of activated carbon have been showed as Fig. 1. With the increase of steam flow, activation time and activation temperature, the yield of ACs decreased, and the iodine value showed a maximum value. This is related to that the yield was only affected by the extent of solution loss and ablation of charcoal. The greater of the extent, the lower of the yield. However, specific surface area, the pore shape and pore volume would of the ACs maybe lead to a lower iodine value.

The formation process of ACs was consisted of trepanning and bearizing. The iodine value has been increased during the trepanning stage. It can be illustrated by the stages of steam flow of 30-40 $g \cdot h^{-1}$ (Fig. 1(a)), activation time of 60-90 minutes (Fig. 1(b)) and activation temperature of 600-750 °C (Fig. 1(c)). Briefly, improved the pore parameters (number, diameter, volume and specific surface area) can brought more sufficient contact between steam and samples, and then deepened the activation level, as well as the faster reaction rate. During the bearizing stage, pore number remained constant, pore volume and specific surface area became lager. Nevertheless, excessive bearizing might make the pore wall thinner and even skeleton destruction occurs. These actually leading to the decrease in numbers and specific surface area of pore, which can be seen in the steam flow over 40 $g \cdot h^{-1}$ (Fig. 1(a)), activation time over 90 minutes (Fig. 1(b)) and activation temperature over 750 °C (Fig. 1(c)).

Taking iodine value and yield of activated carbon into account, the optimal parameters of steam activation were steam flow of 40 g/L, activation time of 90 minutes and activation temperature of 750 °C.

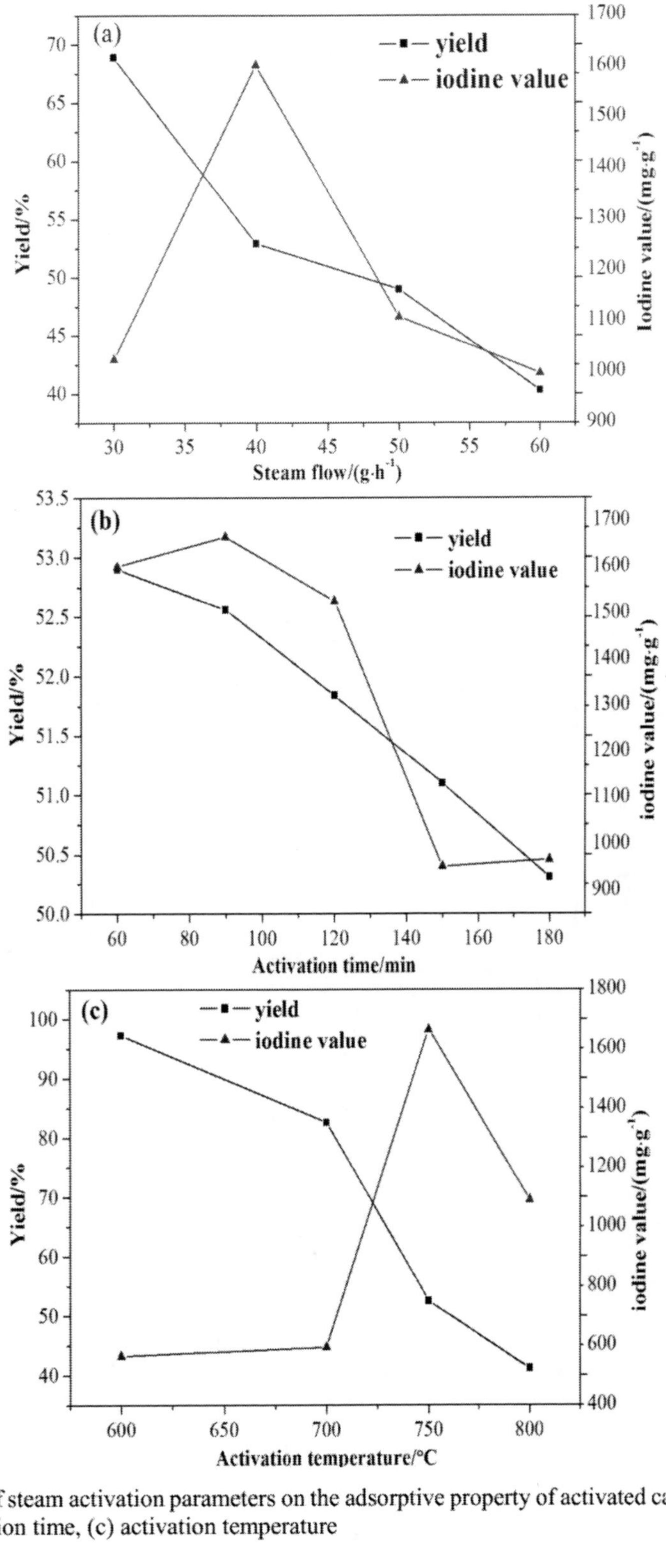

Fig. 1. Effects of steam activation parameters on the adsorptive property of activated carbons: (a) steam flow, (b) activation time, (c) activation temperature

3.2 Effects of alkali activation parameter on the adsorptive property of activated carbon

Single factor experiment was applied to study the effects of immersion time, alkali-carbon ratio, activation temperature and activation time of KOH activation on the yield and iodine value of activated carbon, and have been shown in Fig. 2.

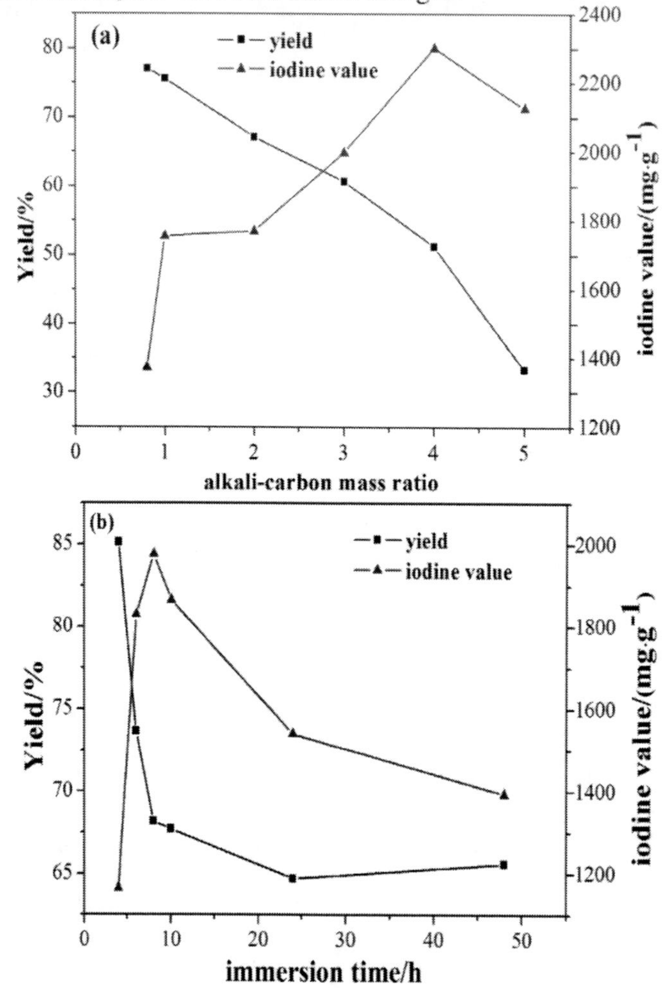

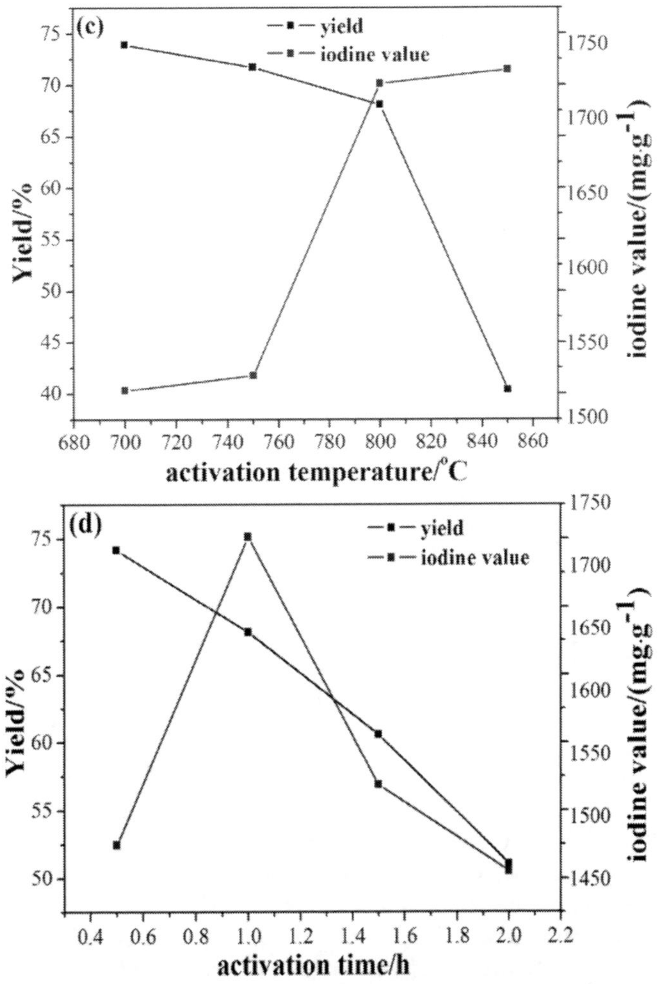

Fig. 2. Effects of KOH activation parameters on the adsorptive property of activated carbon: (a) alkali-carbon mass ratio, (b) immersion time, (c) activation temperature, (d) activation time

Either increasing alkali-carbon ratio or immersion time would expedite the bearizing stage (Fig. 1(a) and (b)). Briefly, the carbon skeletons which around the micorpore will continue react with activator and lead to the gradual collapse of the carbon skeleton. It also increased the number of mesopore and macropore, decrease in specific surface area, and decrease in iodine value as well.

Fig. 2(c) implicated the significant influence of activation temperature on the absorption ability of activated carbon. With the increase of activation temperature, the iodine value increased while yield decreased progressively (750 °C to 800 °C), and then dramatically decreased with the continuing activation (800 °C to 850 °C). This phenomenon can be translated by the chemical properties of K. Briefly, The K has an melting point of 63.2 °C, boiling point of 765.5 °C and density of 0.86 $g \cdot cm^{-3}$, and this properties resulted in the reaction between KOH activator and carbon particles was restricted on the surface (the oxygen-containing groups) during the lower temperature stage. This step was corresponding to the activation-induced stage, and belonging to the pre-activating stage of pore-forming and activation. During the high temperature activation stage, K was presented as liquid which can facilitating flowed in the micropores and reacted with more active sites. Hence, elevating temperature is conducive to formation of micropore inside the carbon particle. However, excessive activation temperature will increase the production cost and machine wear. With the opposite, the yield was decreased with the increase of activation temperature, because the content of destroyed carbon skeleton was increased.

Iodine value reached the maximum when the activated time holding for 60 minutes (Fig. 2(d)). Prolonging activation time, part of micropore may expand to mesopore or macropore, the K steam and other volatile gas might result in attenuation or burnthrough of pore wall which leading to the corrosion of carbon skeleton. It also destructed the pore structure and declined in the absorption ability of ACs. Similarly, with the carbon was ablated continually, the yield was persistently decreased.

Taking iodine value and yield of activated carbon into account, the optimal parameters of KOH activation were alkali-carbon ratio of 4, immersion time of 8 hours, activation temperature of 800 °C and activation time of 60 minutes.

3.3 Characteristic of activated carbon

In order to compare the microstructure of activated carbons from different activation method, RM, CP, AC-1(the activated carbon which prepared with the optimal KOH activation method), and AC-2 (the activated carbon which prepared in the optimal steam activation method) were tested by FTIR, XRD, specific surface, volume and diameter of pore, respectively.

3.4 FTIR and XRD analysis of ACs

FTIR patterns of activated carbons and carbon precursor were showed in Fig. 3(a). Compared with raw material, most of absorption peaks and abundance of carbide reduced dramatically after carbonization. Such as alkane C-H stretching vibration (3000~2850 cm^{-1}), aromatic ring C-H stretching vibration (3100~3000 cm^{-1}), and C-H out-of-plane bending (880~680 cm^{-1}) were disappeared. Only C=C skeletal vibration (1600~1450 cm^{-1}) can be found in the four samples. It have been indicated that carbonization and activation treatment can not only enhanced the carbon skeletal strength and aromaticity, but also made non-carbon elements into small molecules to escape at higher temperature. Compared with AC-1, the abundance of phenol and ketone structure near 3443 cm^{-1}, 1645 cm^{-1}, 1593 cm^{-1}, 1473 cm^{-1}, 1445 cm^{-1} were relatively larger in AC-2, and it fit for the mechanism of steam activation.

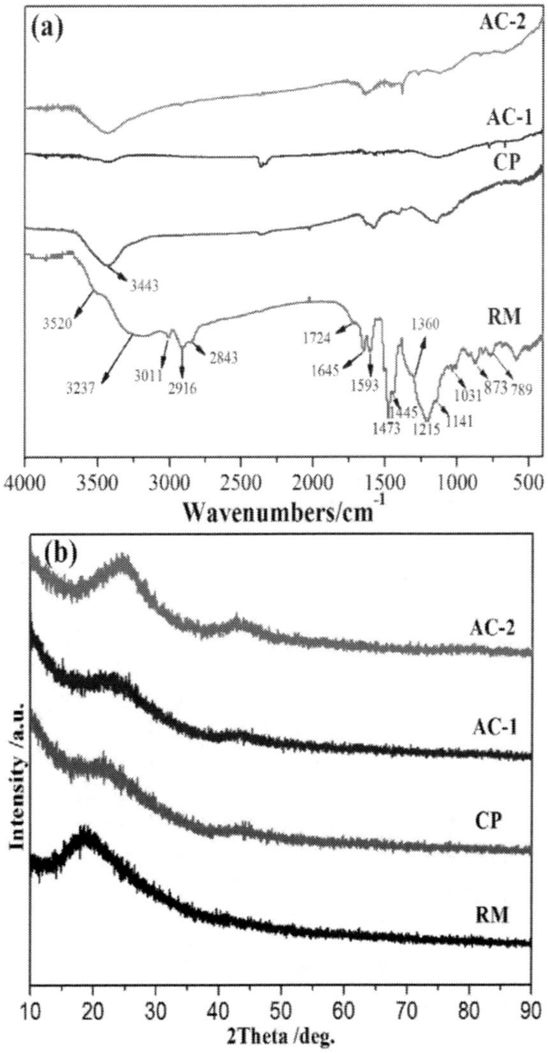

Fig. 3. FTIR (a) and XRD (b) patterns of the samples

As shown in the XRD patterns of ACs and carbon precursor (Fig. 3(b)), the characteristic diffraction peaks of (002) and (100) planes of multilayer graphite structure appeared when 2θ at around 25° or 43°. The diffraction peaks of ACs crystal planes (002) and (100) were wider and the strength was lower than CP. It indicated that the carbon crystalline structure trends to amorphous carbon during the activation process. This phenomenon was more obvious in AC-2 than in AC-1.

3.5 Specific surface area and pore diameter analysis of ACs.

The specific surface area and pore diameter of the activated carbon samples were obtained by nitrogen absorption-desorption isotherm determined at 77 K. According to the adsorption isotherm, the total specific surface area was calculated based on BET method, the pore diameter was calculated based on t-plot method, the distribution of pore size and mesopore volume were calculated based on Barrette Joynere Halenda (BJH) method, the total pore volume was calculated using the N_2 absorption value at the relative pressure (P/P_0) of 0.99. The results have been shown in the table 1.

Table 1. BET surface area and pore structure parameters of activated carbon

Sample	S_{BET}(m²/g)	V_t(cm³/g)	V_{mic}(cm³/g)	V_{mes}(cm³/g)	V_{mes}/V_t(%)	D_p(nm)
AC-1	1293.45	0.708	0.40	0.297	41.95	2.4
AC-2	786.82	0.70	0.13	0.60	85.71	3.54

note: S_{BET} — BET specific surface area, V_t — pore volume, V_{mic} — micorpore volume based on t-plot method, V_{mes} — mesopore volume based on BJH method, D_p — pore diameter

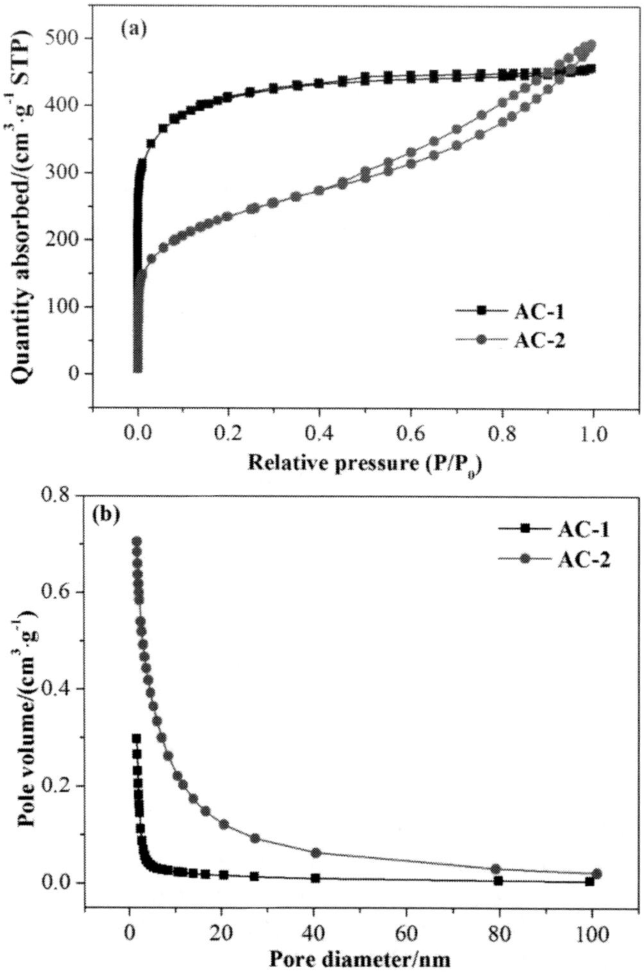

Fig. 4. N_2 adsorption-desorption (a) and BJH adsorption pore size distribution (b) of the ACs

According to the IUPAC classification, the adsorption-desorption isotherm (Fig. 4(a)) of AC-1was basically match the type I, AC-2 exhibited type II, and hysteresis loop was basically match the type H_3 hysteresis patterns. The absorption value of both ACs was increased dramatically with the increase of relative pressure (P/P_0) in the beginning. This phenomenon was corresponding to the filling stage of micropore by N_2. And then the increase tendency became gentle, which was corresponding to the multilayer adsorption of coarse mesopore and macropore. In addition, the condensation occurred inside the pore made a deviation between the isothermal adsorption and desorption isotherm. Since the rising tendency of AC-2 was higher than that of AC-2, it means that the AC-2 had a major proportion of mesopore and macropore than those of AC-1.

It can be seen from the BJH adsorption distribution of pore size from Fig. 4(b). Most of pore size in AC-1 was less than 2 nm, and coexisted with some mesopore with the size of 2-5 nm. AC-1 has a uniform and concentrated distribution of pore size. AC-2 was mainly made up of mesopore, so the N_2 absorption equilibrium was achieved quickly, and in line with the adsorption isotherm. The distribution of pore size on AC-2 has a characterized trailing unimodal. It illustrated that AC-2 has an unevenly distribution of mesopore structure between 2nm and 20nm, and contain a small number of macropore.

4 Summary

The waste external thermal-insulating phenolic foam boards were a kind of high level raw material to produce activated carbon. Both of activation methods and activation conditions have significantly influences on the absorption property of activated carbon. AC-1 was a kind of high quality activated carbon which mainly composed of micropore and mesopore (the iodine value was 2300 $mg \cdot g^{-1}$, specific surface area 1295 $m^2 \cdot g^{-1}$, average pore size 2.4 nm) were activated by KOH with the ratio of alkali/carbon was 4:1, immersion time of 10 hours, activation temperature of 800 °C and activation time of 40 minutes. The AC-2 was mainly composed of 2-20 nm mesopore, the iodine value was 1665 $mg \cdot g^{-1}$, specific surface area was 787 $m^2 \cdot g^{-1}$. AC-2 was prepared by steam activation at the temperature of 750 °C for 90 minutes, and the steam flow was 40 $g \cdot h^{-1}$.

This work was supported by the National Natural Science Foundation of China (U1361126), the Specialized Research Fund for the Doctoral Program of Higher Education (20132120110001), the Young Teacher Foundation of University of Science and Technology Liao Ning (QN201625).

References

1. J. Mao, J. Chang, Y. Chen, D. Fang, Chemical Industry and Engineering, **15** (1998) 38.
2. H. Shen, A. J. Lavoie, S. R. Nutt, Composites Part A: Applied Science and Manufacturing, **34** (2003) 941.
3. D. Lozano-Castello, M. Lillo-Rodenas, D. Cazorla-Amoros, A. Linares-Solano, Carbon, **39** (2001) 741.
4. B. Acevedo, C. Barriocanal, I. Lupul, G. Gryglewicz, Fuel, **151** (2015) 83.
5. D. Lozano-Castello, D. Cazorla-Amoros, A. Linares-Solano, Energy & Fuels, **16** (2002) 1321.
6. H. Teng, H. c. Lin, AIChE Journal, **44** (1998) 1170.
7. J. i. Hayashi, A. Kazehaya, K. Muroyama, A. P. Watkinson, Carbon, **38** (2000) 1873.
8. M. A. Islam, I. Tan, A. Benhouria, M. Asif, B. Hameed, Chemical Engineering Journal. **270** (2015) 187.
9. M. A. Islam, A. Benhouria, M. Asif, B. Hameed, Journal of the Taiwan Institute of Chemical Engineers (2015).
10. M. Munoz-Guillena, M. Illán-Gómez, J. Martin-Martinez, A. Linares-Solano, C. Salinas-Martinez de Lecea, Energy & Fuels, **6** (1992) 9.
11. M. Greenbank, S. Spotts, Water Technol, **16** (1993) 56.
12. P. Hadi, M. Xu, C. Ning, C. S. K. Lin, G. McKay, Chemical Engineering Journal, **260** (2015) 895.

13. I. Okman, S. Karagöz, T. Tay, M. Erdem, Applied Surface Science, **293** (2014) 138.
14. A. Elmouwahidi, Z. Zapata-Benabithe, F. Carrasco-Marín, C. Moreno-Castilla, Bioresource Technology, **111** (2012) 185.
15. C. Peng, X.-b. Yan, R.-t. Wang, J.-w. Lang, Y.-j. Ou, Q.-j. Xue, Electrochimica Acta, **87** (2013) 401.

Exploiting a Simple Method for the Determination of Manganese in Polyethylene Lined Tubing for Petroleum and Natural Gas Industries

Xiaodong Shao [1,2*], and *Xin* Shi[3]

[1]State Key Laboratory for Performance and Structure Safety of Petroleum Tubular Goods and Equipment Materials, No. 89 Jinyeer Road, Xi'an 710077, China
[2]CNPC Tubular Goods Research Institute, No. 89 Jinyeer Road, Xi'an 710077, China
[3]Engineering Institute of Technology, Sinopec Northwest Oilfield Branch, Urumqi 830011, China

Abstract. The polyethylene lined tubing is the key to enabling the industry to meet some of the energy security challenges that nations face today. It is well known that manganese is an important element in polyethylene lined tubing. In this paper, a simple spectrophotometric method was described for the determination of manganese in polyethylene lined tubing. The method was based on the oxidation-reduction reaction between ammonium persulfate and manganese(II) producing manganese(VII) in the presence of silver nitrate as a catalyst. The characteristic wavelength of maximum absorption of manganese(VII) was obtained locating at 530 nm. Under the optimum reaction conditions the absorption value was proportional to the concentration of manganese in the range of 0.2%~1.9% ($R^2 = 0.9997$), and the relative standard deviation was less than 3.0% (n=5). The proposed method was applied successfully to determine manganese in polyethylene lined tubing real samples.

1 Introduction

In the west of China, the area has rich petroleum resource, which has become an important exploitation target now. The condition of gas and oil wells in the west of China has complex characteristics such as deep, high temperature, high pressure, and serious corrosive that only corrosion-resistant tubing can be applied to explore. One key feature there is obvious existence of corrosive

Fig. 1. Polyethylene lined tubing samples

* Corresponding author: shaoxd@163.com

hydrogen sulfide medium. The sulfide stress corrosion failure frequently occurs due to absence of applicable technical specification which has led to huge economic damage. Statistical data showed that the average loss is 100 thousand dollar for the tubing failure. In some areas no suitable tubing can be supplied due to high hydrogen sulfide content. So there is obvious safety risk when tubing in these gas fields. Specification for tubing used in these regions was very necessary and urgent. The polyethylene lined tubing (Fig. 1) has excellent resistance to corrosion. These composite pipes are key to enabling the industry to meet some of the energy security challenges that nations face today. Thus, polyethylene lined tubing will be widely used in the exploitation of acid gas and oil wells. The content of chemical elements in polyethylene lined tubing directly influences the properties of materials. In order to control the quality of polyethylene lined tubing products, the contents of trace impurities and alloy elements in polyethylene lined tubing should be determined. In making processes, the content of alloyed elements such as manganese, nickel, chromium, molybdenum, vanadium, and so on, is required to be strictly controlled because it fundamentally determines the performance of polyethylene lined tubing.

It is well known that manganese is an important element in polyethylene lined tubing. Therefore, a rapid and precise analytical method for the determination of manganese is essentially required for the production control of polyethylene lined tubing. Manganese is valuable alloying agents and coexists in a number of industrially important alloys and steels. Owing to their increasing industrial applications of manganese, the determination of trace amounts of manganese is of great interest. Therefore, it was important from the analytical point of view to develop simple, sensitive and reliable methods for the determination of manganese in various complex matrixes. Numerous analytical methods were employed for the determination of manganese in different types of matrices including atomic emission spectrophotometry [1,2], atomic absorption spectrometry [3], laser induced breakdown spectroscopy [4], atomic fluorescence spectroscopy [5], neutron activation analysis [6], electroanalytical method [7] and spectrophotometry [8]. Spectrophotometric methods occupy special position due to their simplicity, less expensive instrumentation and high sensitivity. Similarly, spectrophotometric methods also play an important role in manganese determination. The literatures shows that aromatic amines, azo dyes, oximes, porphyrins and other reagents have been used for the colorimetric determination of manganese. The most significant reagents are permanganate, formaldoxime and 1-(2-pyridylazonaphthol), which is far more sensitive than the other two reagents. This present method was based on the oxidation-reduction reaction between ammonium persulfate and manganese(II). Employing ammonium persulfate as an oxidant, manganese(II) was oxidized to manganese(VII) in the presence of silver nitrate as a catalyst at the room temperature. The characteristic wavelength of maximum absorption was measured at 530 nm. Under the optimum reaction conditions the absorption value (A) was proportional to the concentration of manganese in the range of 0.2%~1.9% (mass fraction), and the relative standard deviation (RSD) was less than 3.0% (n=5). The proposed method was free from the interference from a large number of analytical important elements and has been applied satisfactorily to the determination of manganese in polyethylene lined tubing samples with the improved accuracy and precision. Results were in agreement with those attained by inductively coupled plasma atomic emission spectrophotometry at the 95% confidence level. The satisfactory performance in the determination of manganese in polyethylene lined tubing demonstrated that the method was practical and suitable not only for quality control analysis but also for product analysis, confirming the promise for polyethylene lined tubing research.

2 Experiment

2.1 Apparatus

A UV-visible spectrophotometer (Model TU-1901, Beijing Persee General Instrument Co. Ltd.) with 1.0-cm quartz cells was employed for all absorbance measurements.

2.2 General procedures

Weigh accurately 0.1000 g of drilled low alloy steel samples and dissolve the samples in 20 mL mixture acid solution of sulphuric acid 15% (V/V) and phosphoric acid 15% (V/V) in a 150-mL taper bottle. Continue the treatment with mixture acid till the sample goes into solution and heat to fumes of oxides of nitrogen, sulphuric acid and evaporate almost to dryness. Cool the solution, add 50 mL deionised water to dissolve some materials. Then, add 5 mL of 5.0 mg·mL^{-1} solution of silver nitrate, 10 mL of 200.0 mg·mL^{-1} solution of ammonium persulfate in the taper bottle to oxidate manganese(II) to manganese(VII) oxidation states. Boil the solution to no bubble and cool the solution at room temperature. Finally, dilute the solution in a 100-mL volumetric flask, followed make up to the mark with deionised water, mix well and let stand for 10 min. Then, transfer the above mulberry stock solution in 1.0-cm quartz cells. Add 5.0 mg·mL^{-1} solution of sodium nitrite to the remaining solution reduce manganese(VII) to the lower oxidation states, this transparent solution was employed as the blank solution in subsequent experiments. Measure the absorbance at 530 nm using a UV-visible spectrophotometer with 1.0-cm quartz cells against a reagent blank prepared similarly. Plot the amount of manganese in the sample solution against absorbance to obtain the calibration graph.

3 Results and discussion

3.1 Spectral characteristics

The absorption spectra of the reagents blank, manganese(VII) against the reagent blank were tested from 400 nm to 700 nm. The manganese(VII) yields the characteristic wavelength of maximum absorption at 530 nm wavelength. However, the absorption spectrum of the reagents blank under similar conditions shown that it does not absorb at 530 nm. And the experimentation also demonstrated that manganese(VII) increases the absorbance value considerably, resulting in increased sensitivity of the method. Therefore, measure the absorbance of manganese(VII) at 530 nm.

3.2 Select dissolved acid for samples

The effect of the different acids for dissolve the low alloy steel samples was examined. It was found that a solution mixed with 15% (V/V) sulphuric acid and 15% (V/V) phosphoric acid has a good effect for dissolve the low alloy steel samples. Thus, an acid solution mixed with 15% (V/V) sulphuric acid and 15% (V/V) phosphoric acid was recommended for dissolve the low alloy steel samples in subsequent studies.

3.3 Influence of the amounts of oxidant

Ammonium persulfate was selected as the oxidant in the proposed method. The effect of the oxidant was examined by measuring the absorbance of solution containing certain amounts of manganese(II) and variable amounts of ammonium persulfate. It was found that 10 mL of 200.0 mg·mL^{-1} solution of ammonium persulfate sufficed to oxidate the amounts of manganese(II) taken, with higher concentrations the absorbance was essentially constant. Ten millilitres of 200.0 mg·mL^{-1} ammonium persulfate solution were recommended as a suitable amount of oxidant.

3.4 Influence of the amounts of silver nitrate

The oxidation-reduction reaction between manganese(II) and ammonium persulfate without a catalyst was slowly and not fully. Selected the silver nitrate as a catalyst for the oxidation-reduction reaction between manganese(II) and ammonium persulfate. It has been further observed that the oxidation-reduction reaction between manganese(II) and ammonium persulfate occurs immediately even at room temperature if silver nitrate was used. Therefore, 5 mL of 5.0 mg·mL^{-1} solution of silver nitrate was employed in subsequent experiments.

3.5 Influence of the reductant

Sodium nitrite was selected as the reductant in the proposed method. 5.0 mg·mL^{-1} solution of sodium nitrite reduce manganese(VII) to the lower oxidation states manganese(II). The transparent solution was employed as the blank solution in subsequent experiments.

3.6 Sequence of addition of reagents

From experiments in which sulphuric acid, phosphoric acid, ammonium persulfate and silver nitrate were added in all possible orders. It was found that there was no appreciable change in the colour of the solution and the variations in the absorbance values. These phenomenons demonstrate that the oxidation-reduction reaction between manganese(II) and ammonium persulfate was independent of the order of addition.

3.7 Preparation of the calibration graph

Under the optimal conditions, a series of standard solutions of manganese were prepared by the different steel certified reference materials in a 100-mL volumetric flask. Add 5 mL of 5.0 mg·mL^{-1} solution of silver nitrate, 10 mL of 200.0 mg·mL^{-1} solution of ammonium persulfate followed make up to the mark with deionised water, mix well and let stand for 10 min. Measure the absorbance at 530 nm using a UV-visible spectrophotometer with 1.0-cm quartz cells against a reagent blank. The absorbance of the solution was proportional to the concentration of manganese in the range of 0.2%~1.9% (mass fraction). Plot the amount of manganese in the sample solution against absorbance to obtain the calibration graph. The linear regression equation for manganese was $A = 0.41837C - 0.01073$, $R^2 = 0.9997$, and the relative standard deviation (RSD) was less than 3.0% (n=5).

3.8 Interference studies

The extent of interference by diverse ions was determined by measuring the absorbance of solutions containing manganese and various amounts of diverse ions. The criterion for interference was an absorbance value varying by more than +2% from the expected value of manganese alone. The results show that a large excess of cations and anions which are usually associated in the determination of manganese do not interfere. The colorless metal ions do not interfere. The tolerable amounts of foreign species with respect to manganese for interference at +2% level were over 200~300 for the colored metal ions. When the colored metal ions contents of not more than 200~300 times manganese, it can be in some color solution adding a few drops of EDTA solution or sodium nitrite solution to fade the color, this solution as the measured reference solution to eliminate the interference. There was not obvious interference ions exist in polyethylene lined tubing samples.

4 Applications

4.1 Determination of manganese in polyethylene lined tubing samples

The recommended procedure has been applied satisfactorily to the determination of manganese in polyethylene lined tubing samples. Measure the absorbance at 530 nm using a UV-visible spectrophotometer with 1.0-cm quartz cells against a reagent blank. The results were summarized in Table 1 with the good accuracy and precision. To verify the results obtained by the proposed method, inductively coupled plasma atomic emission spectrophotometry method was applied to determine the samples according to the literature and the results obtained by the proposed method agreed well with that from the reference method.

Table 1. Results of determination of manganese in polyethylene lined tubing samples[a]

Sample No.	Size (mm)	*A*	RSD (%)	Content (ω, %)	
				By the proposed method	By AES
1	Φ73	0.2570	2.51	0.64	0.65
2	Φ73	0.2545	1.87	0.63	0.63
3	Φ73	0.5289	0.96	1.29	1.32
4	Φ73	0.5373	1.09	1.31	1.30
5	Φ73	0.5457	2.16	1.33	1.33
6	Φ73	0.5444	1.12	1.33	1.35
7	Φ73	0.5708	0.66	1.39	1.41
8	Φ73	0.5624	0.79	1.37	1.35
9	Φ89	0.1943	2.77	0.49	0.51
10	Φ89	0.2068	2.25	0.52	0.50
11	Φ89	0.5917	1.53	1.44	1.46
12	Φ89	0.6043	1.24	1.47	1.48
13	Φ89	0.4829	0.92	1.18	1.16
14	Φ89	0.4913	0.30	1.20	1.20
15	Φ89	0.2236	1.59	0.56	0.56
16	Φ89	0.2213	2.00	0.55	0.57

Notes: [a]The average of 5 determinations.

5 Conclusions

Based on the oxidation-reduction reaction between manganese(II) and ammonium persulfate, a simple and accurate spectrophotometric method for determination of manganese in polyethylene lined tubing was proposed. The purple-red colored manganese(VII) solution exhibits absorption maximum at 530 nm. The proposed method was free from the interference from a large number of analytical important elements and has been applied satisfactorily to the determination of manganese in polyethylene lined tubing samples. Compared with other methods for the determination of manganese, this method offers advantages of simplicity, less expensive instrumentation and high efficient. The satisfactory performance in the determination of manganese in polyethylene lined tubing demonstrated that the method was practical and suitable not only for quality control analysis but also for product analysis, confirming the promise for polyethylene lined tubing research.

The authors gratefully acknowledge the State Key Laboratory for Performance and Structure Safety of Petroleum Tubular Goods and Equipment Materials, and CNPC Tubular Goods Research Institute.

References

1. W. Li, P. Simmons, D. Shrader, T.J. Herrman, S.Y. Dai, Talanta **112,** 43 (2013)
2. N. Ozbek, S. Akman, Food Chemistry **200,** 245 (2016)
3. S.M.A. Azeem, S.M.M. Attaf, M.F. El-Shahat, Reactive and Functional Polymers **73,** 182 (2013)
4. M.R. Pourjavid, A.A. Sehat, Microchemical Journal **129,** 259 (2016)
5. S. Feng, D. Yuan, T. Zhou, Analytica Chimica Acta **963,** 53 (2017)
6. K. Tomura, H. Tomuro, Journal of Radioanalytical and Nuclear Chemistry **242,** 147 (1999)
7. R. Piech, B. Baś, W.W. Kubiaka, Journal of Electroanalytical Chemistry **621,** 43 (2008)
8. S. Feng, Y. Huang, T. Zhou, Continental Shelf Research **92,** 37 (2015)

MATEC Web of Conferences **142**, 01007 (2018) https://doi.org/10.1051/matecconf/201814201007
ICMAE2017

Failure Analysis of Cathode Materials for Energy Storage Batteries in Overcharge Test

Hongwei Wang[1,*], *Ziqiang* Tao[1], *Nianpeng* Si[2], *Yanling* Fu[1], *Tao* Li[1] and *Haiqing* Xiao[1]

[1] Institute of Industrial and Consumer Product Safety, Chinese Academy of Inspection and Quarantine, Beijing Petrochemical Engineering Co., Ltd., Beijing, China
[2] Inspection & Quarantine Technology Center of Guangdong Entry-Exit Inspection & Quarantine Bureau of P. R. China, Room 812, Tower B, No.66 Huacheng Avenue, Guangzhou, Guangdong Province, China

Abstract. The micro-analysis of energy storage batteries in overcharge test at 20℃ temperature was investigated. The results showed as follows: (1) Compared with the normal battery charge at room temperature scanning microscope maps of battery overcharge, the crystal was fractured of the positive surface when the battery was overcharging, and with the increase of experimental time, the crystal fracture is more serious, this is to say that the battery electrolyte decomposition or battery material structure could change by overcharge test. (2) Based on the XRD maps, the characteristic peak position of overcharge test of cathode materials were changed, this showed that the content of cathode material elemental has been destroyed and changed in overcharge test.

1 Introduction

It is considered that Lithium-ion battery is electric vehicle power supply important development direction the next few years, because the features of Lithium-ion battery is excellent, such as high working voltage, light weight, high Lithium-ion battery specific energy, long cycle life, fast charging, and so on[1-3]. Lithium-ion battery has been more and more widely used in mobile electronic equipment, electric vehicles, as well as national defense and other high-tech industry. Retired lithium-ion battery is often used in the field of energy storage[4-5].

Lithium-ion batteries are usually used constant current constant voltage charging mode. Charging begins to constant current stage, the battery voltage is low, and the charging current is stable in the charge process. As the charge continues, the battery voltage gradually increased to 4.2V, and the charger should be immediately transferred to the constant voltage charging, charging voltage fluctuations should be controlled within 1%, the charging current gradually reduced. When the current drops to a certain range, into the trickle charge stage, and finally make the battery in a sufficient state. The battery overcharge is continued to charge after a certain charging process. The damage of the lithium battery overcharge is great, and it may be fire or explosion.

* Corresponding author: author@email.org

Therefore, the microcosmic change of the positive electrode after the overcharge of the lithium energy storage battery is analyzed, and the change rules of the positive electrode is studied, which provides some basic research data for the improvement of the performance of the energy storage battery.

2 The overcharge tests

2.1 Test sample

The parameters of sample battery in this paper as follows: 1) The model: domestic battery, rated voltage 3.7 V, rated capacity 10 Ah, internal resistance < 6 mΩ, weight < 320 g; 2) The composing of the battery: the cathode material is $LiMn_2O_4$, the anode material is graphite, the electrolyte is $LiPF_6$, EC and DMC, and battery separator is celgard 2325; 3) Dimensions: 133 mm×66 mm×18 mm.

2.2 Test instrument

Maccor battery performance test systems (model M4200), CSZ Thermostat box (model ZPS-16SPHH-101), Bruker D8 focus X instrument and Hitachi S-4800 instrument. The test is shown in Fig. 1.

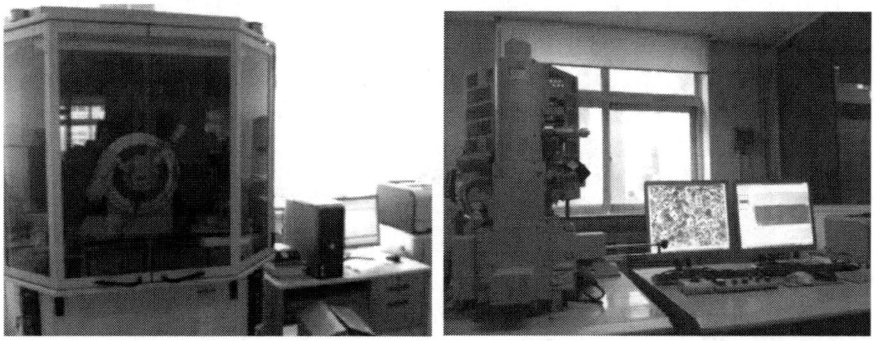

Fig. 1. The overcharge tests.

2.3 Test method

1) Each test battery was discharged at a constant current of 2.2 A, to a final discharge voltage 2.7 V, was charged at constant current of 2.2 A to the 4.2 V, and was charged at constant voltage to 0.2 A. the break of battery charge and discharge is 1 hour. The 10 cycle tests were carried out.

2) The test was carried out in an ambient temperature of 20 ℃±5 ℃. Each test battery was discharged at a constant current of 2.2 A, to a final discharge voltage 2.7 V that was specified by the manufacturer. Then, Sample batteries was be charged at a constant current of 2.2 A to 5 V.

3 Results and discussion

The batteries of rated voltage 3.7V were used in this paper, charge terminal voltage is 4.2V, and the allowable voltage error range is 0.042V. Figure 2 is the positive of the overcharge battery in accordance with the above method.

Fig. 2. The positive of the overcharge battery.

SEM morphology of positive electrode of battery in charge and overcharge test is shown in Fig.3 and Fig.4 (magnification of 9000). Energy spectrum element content table of positive after overcharge test is shown in Table1.

Fig. 3. SEM morphology of positive electrode of battery in charge test.

Fig. 4. SEM morphology of positive electrode of battery in overcharge test.

Table 1. Energy spectrum element content table of positive.

	Element	Weight percentage /%	Atomic percentage /%
Overcharge	C	364.67	20.80
	O	424.78	18.19
	F	1448.01	52.22
	P	107.85	2.39
Charge	C	832.44	69.63
	O	250.55	15.73
	F	239.61	12.67
	P	34.49	1.12

According to Fig.3, Fig.4 and Table 1, the positive structure of battery has been crystallized in overcharge test, and different degrees fracture has been occurred. Compared with SEM morphology of the normal charge and overcharge battery at room temperature, the crystal was fractured of the positive surface when the battery was overcharging, and with the increase of experimental time, the crystal fracture is more serious, this is to say that the battery electrolyte decomposition or battery material structure could change by overcharge test.

Fig. 5 shows the X-ray diffraction image of the positive electrode of the battery after normal charge and overcharge test.

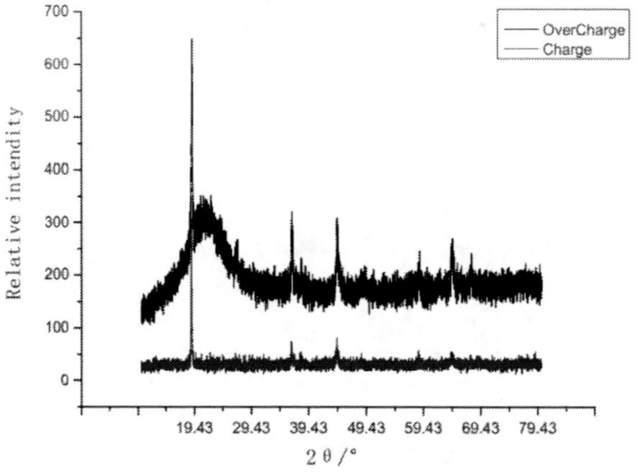

Fig. 5. XRD pattern of positive electrode of battery in charge and overcharge test.

It can be seen from Fig. 5, compared to the normal state of the charge battery, the content and type of positive elements of battery are higher than in overcharge test, overcharge lead to the element content of the battery positive material has been destroyed and changed.

Therefore, the negative capacity than the positive capacity is higher according to the battery design, the gas generated from positive go through the diaphragm complex with cadmium that from negative, and the battery pressure will not be significantly increased under normal circumstances, but if the charging current is too large, or charging time is too long, the oxygen is too late to be consumed, it may cause internal pressure, battery deformation, and other undesirable phenomena. At the same time, its electrical performance will be significantly reduced.

4 Conclusions

On the basis of the foregoing test results, analysis and discussions, the conclusion can be obtained as follows:

(1) Compared with the normal battery charge at room temperature scanning microscope maps of battery overcharge, the crystal was fractured of the positive surface when the battery was overcharging, and with the increase of experimental time, the crystal fracture is more serious, this is to say that the battery electrolyte decomposition or battery material structure could change by overcharge test.

(2) Based on the XRD maps, the characteristic peak position of overcharge test of cathode materials were changed, this showed that the content of cathode material elemental has been destroyed and changed in overcharge test.

The authors would like to acknowledge the financial support provided for this research work by Fund: Fundamental scientific research projects of Chinese Academy of Inspection and Quarantine (2017JK024).

References

1. H. YU,H. WANG,Y. WANG. et al. PTCA(PART A: PHYS.TEST.) **49**: 535-539(2013)
2. H. YU,H. WANG,Y. Fu. et al. *2014 International Conference on Mechanics and Materials Engineering*. 673-679(2014)
3. H. Wang, W. Wu, W. Li. et al. Chinese Journal of Power Sources. **38**(8): 1430-1432(2014)
4. L.Liu, S. Tian, Y. Zhu, W. Tang, L.L. Li, Y.P. Wu, J. Chin. Chem. Soc., **59**, 1216-1219 (2012).
5. W. Tang, L. Liu,S. Tian. et al. Electrochemistry Communications **13**, 1159-1162(2011)

Effect of Ag Doping on the Electronic Structure and Optical Properties of ZnO(0001) Surface

Qian Xiang[1] *, *Shutao* Zhao[1] , *Yanning* Wu[1], and *Guangdong* Liu[1]

[1]School of Physics and Electronic Engineering, Fuyang Normal College, Fuyang 236037, China

Abstract. Using first-principle calculations, the geometrical structure, the electronic and optical properties of Ag-doped ZnO(0001) surface have been investigated. We found that Ag-doped ZnO(0001) surface is more easily formed on the first layer. On the other hand, the doped surface has gradually become an equipotential body, showing obvious metallic characteristics. We found that a new peak appeared in the low energy region after Ag doping, which was mainly due to the electron transition between the two orbital levels of Ag-4d and O-2p.

1 Introduction

ZnO is a direct wide band gap(3.37eV) semiconductor with a large excitation binding energy of 60 meV at room temperature[1]. Thanks to these properties, ZnO is used as a photoelectronic material for solar cells, gas sensors, transistors[2-3]. In order to change its electronic and optical properties, the doping of ZnO surface has been a popular technique, such as F-B[4], Cu[5], Al[6]. Due to the ZnO crystal has a spontaneous polarization direction along the c axis,different polar surfaces exhibit different growth characteristics.[7-9]. Lately, from the experiment, Suat Pat[10] found the ZnO and AZO thin films have high transparency using RF sputtering technique. Long xing Su[11] has reported Be doping lead to ZnO surface gap widening. From the theory, Jianping Xiao[12] has proposed that Cu-doped ZnO(0001) surface promote efficient formation of oxygen vacancies. Haifeng Zhang[13] indicated that Si atoms adsorbed on ZnO (0001) surface lead to a slightly blue shift phenomenon. Mohammed Ali Lahmer[14] has reported that hydrogen adsorption is more favored on Mg doping ZnO(000$\bar{1}$) surface. So far, Ag-doped ZnO(0001) surface has few reports on theoretical studies.

In this paper, geometrical structure, electrical properties, and optical properties of Ag-doped ZnO(0001) surface are systematically studied by first principles calculations.

2 Computational methods

All calculations are performed by the first-principles method based on the density functional theory with the generalized gradient approximation (GGA)[15], and the projector augmented wave (PAW)[16]pseudo-potentials, as implemented in the VASP codes[17, 18].

* Corresponding author: xiangqian8702@163.com

The gamma centred 8 × 8 × 1 k-points is used for the Brillouin zone. The computational model contains five bilayers in which the two bottom ZnO bilayers are fixed while the top three ZnO bilayers and adatoms are allowed to relax. We carry out calculation with 10Å thick vacuum layers. To prevent unphysical charge transfer between the bottom and top slabs, pseudohydrogens are used on the bottom layer[19]. $Zn^3d^{10}4s^2$, $O2s^22p^4$and $Ag4d^{10}5s^1$ are chosen as valence electrons. The computational model is shown in Fig.1.

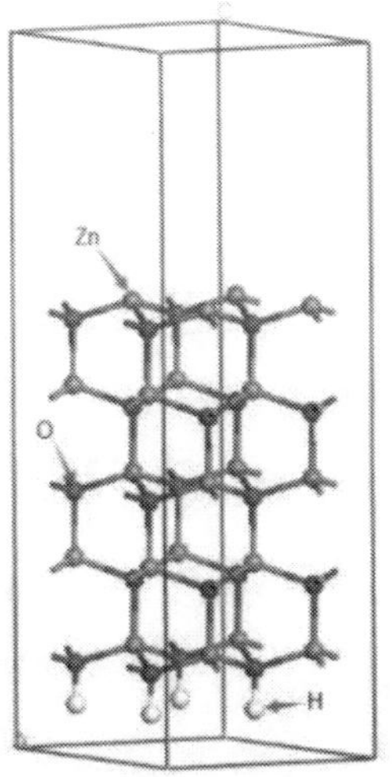

Fig. 1. Configurations of ZnO(0001)surface. Four pseudohydrogen atoms (white ball) are attached to the bottom of the cell

3 Results and discussion

3.1 Geometry structure.

By optimizing the geometry structure of the pure ZnO(0001) surface, The Zn-O bilayer spacings are 0.493Å, 0.591Å and 0.601Å for the first three pairs (counted from the surface), which are slightly smaller than the value in bulk(0.634Å). On the other hand, The Zn-Zn bilayer spacings are 2.711Å, 2.678Å, 2.663Å for the first three pairs (counted from the surface), which are slightly bigger than the value in bulk(2.638Å). This result shows that relaxation only affects the bilayer near the surface, but the relaxation of the deep bilayer is negligible. This result agrees with previous study[20].

3.2 Electrical properties.

The total density of states and projected density of states of Ag-doped ZnO (0001) surface at three layers are presented in Fig.2, respectively. From Fig. 2. We find that the Fermi levels are already in the conduction band, and the Fermi energy gradually moves into the deep conduction band as the doping layer deepened. The surface states near the conduction band minimum(CBM) are mainly contributed by the first layer of the Zn-O bilayer, and the contribution of the second and third layer Zn-O bilayers is less. Therefore, we chose projected density of states (PDOS) of Ag-doped ZnO(0001) surface at first layers to make further analysis. As shown in Fig.2.(b), The valence band maximum(VBM) of Ag-doped

ZnO(0001) surface at first layers is mainly contributed by Ag-4d and O-2p, and the conduction band minimum is mainly contributed by Zn-4s and Ag-5s.

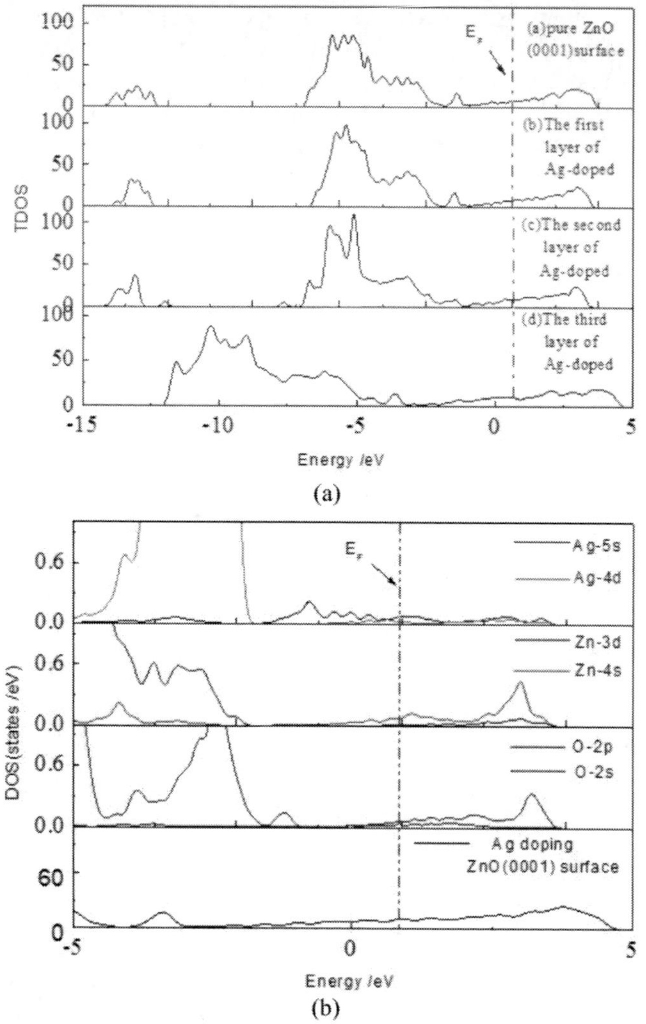

Fig. 2. (a): the total density of states (TDOS) of pure ZnO (0001) surface and the total density of states (TDOS) of Ag-doped ZnO(0001) surface. (b): projected density of states (PDOS) of Ag-doped ZnO(0001) surface at first layers.

In order to study the stability of doped surfaces, the formation energies and electrostatic potential of Ag doped with three layers on ZnO (0001) surface were calculated respectively. The formation energies are as follows: the first layer is -1.65ev; the second layer is -1.39ev; the third layer is -1.29ev. From the change of formation energy, it can be seen that the formation energy of Ag atom doped at the third layer position is highest, and the formation energy decreases gradually as the doping atom position moves to the surface. Therefore, we believe that Ag atoms are more likely to be doped on the first layer of ZnO (0001) surface.

In addition, we calculated the electrostatic potential energy of the pure ZnO (0001) surface and the electrostatic potential energy of the first layer doped Ag on the ZnO (0001) surface along the vertical surface (Z axis), as shown in Fig. 3. The electrostatic potential energy is obtained by the static calculation of the surface model after relaxation. In the calculated surface model, the work function is calculated from the electrostatic potential energy and Fermi energy. The work function is calculated as follows:

$$\phi = x - \varepsilon_F \qquad (1)$$

where ϕ is work function. x is electrostatic potential energy. ε_F is Fermi energy.

From the diagram, it is found that the electrostatic potential of each layer decreases as gradually enters the surface, especially at the first layer of the surface. The electrostatic potential fluctuates greatly, and the electrostatic potential energy is lower at the position near the center of the atom, and between the two atomic layers the value is the maximum. Especially after the first layer on the Ag doped ZnO (0001) surface, the electrostatic potential energy increases obviously. The main reason is that the surface structure changes greatly after surface relaxation. Through calculation, we can get that the work function of ZnO (0001) surface is 2.212ev, and the experimental value is 2.8ev[21].

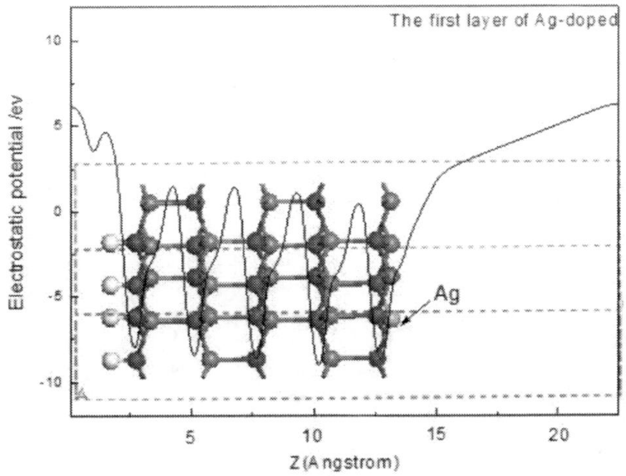

(a)

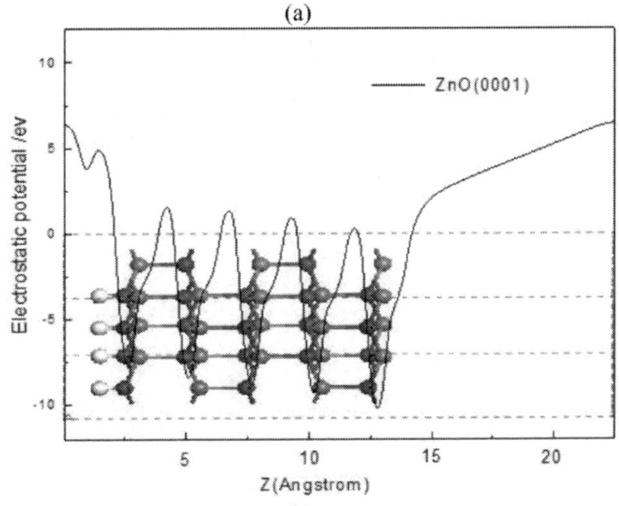

(b)

Fig. 3. (a)the electrostatic potential energy of the first layer doped Ag on the ZnO (0001)surface, (b)The electrostatic potential energy of the pure ZnO(0001) surface

3.3 Optical properties.

Sequentially, in order to reveal the relationship between the electronic structure and optical properties, it is necessary to study the imaginary part of the dielectric function. Fig.4 shows the dielectric function of pure ZnO(0001) surface and the first layer doped Ag on the ZnO

(0001) surface. As we can see from the diagram, the imaginary part of their dielectric function is basically the same. However, the peak value is higher compared to the pure surface at 1.25eV. Since each dielectric function peak corresponds to an absorption maximum, the corresponding absorption photon energy at the 1.25eVenergy is 4.23eV. This is mainly due to the electronic transitions between the two orbital levels of Ag-4d and O-2p.

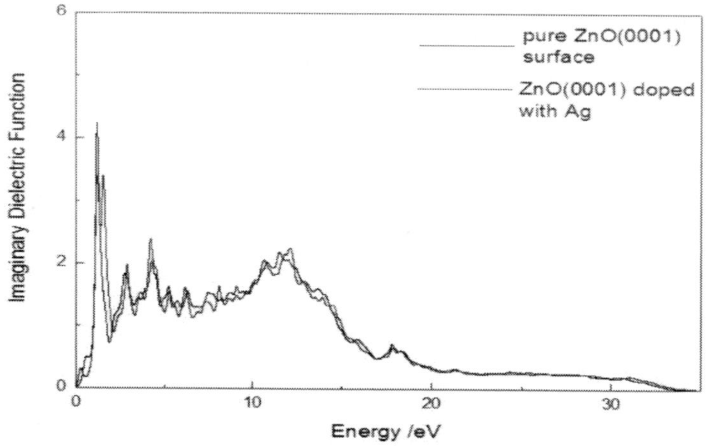

Fig. 4. the dielectric function of pure ZnO(0001) surface and the first layer doped Ag on the ZnO (0001) surface

4 Conclusions

In summary, geometrical structure, electrical properties, and optical properties of Ag-doped ZnO(0001) surface are investigated through the first-principle calculations. The calculated results show that the Fermi energy has entered into the conduction band after doping Ag, and the Fermi energy gradually move into the deep conduction band as the doping layer deepened. At the same time, we found that Ag preferred doping on the first layer of ZnO (0001) surface. From the overall point of view, the doped surface has gradually become an equipotential, showing obvious metallic characteristics. Simultaneity, A new peak is observed at 1.25eV, which is mainly due to the electronic transitions between the two orbital levels of Ag-4d and O-2p. We hope these results could provide useful information about metal atoms doped on ZnO surfaces.

This work was supported by Natural Science Foundation of Fuyang Teachers College (2016FSKJ12), National Natural Science Foundation of China(11604052), Natural Science Foundation of Anhui Province(KJ2017A343), Open subject of State Key Laboratory of Low-Dimensional Quantum Physics(KF201708).

References

1. D. C. Look, B. Claflin, Y. I. Alivov, S. J. Park, Physica Status. Solidi(a).**201**(2004).
2. K.Mahmood, B.S. Swain, Gill-Sang. Han, ACS Appl. Mater. Interfaces, **6**(2014).
3. P.S. Venkatesh, K. Jeganathan, J. Solid State Chem, **200**(2013).
4. Q. Li, L.P. Zhu, Y.G. Li, Journal of Alloys and Compounds, **697**(2017).
5. M.A. Lahmer, Journal of Physics and Chemistry of Solids, **89**(2016).

6. D.M. Ma, Z. Wang, H.T. Cui, Sensors and Actuators B, **224**(2016).
7. H. Xu, K. Ohtani, M. Yamao, H. Ohno, Appl. Phys. Lett, **89**(2006).
8. R. Zhu, Q. Zhao, J. Xu, B.G. Liu, Cryst. Eng. Comm., **14**(2012).
9. Z.L. Wang, X.Y. Kong, J.M. Zuo, Phys. Rev. Lett., **91**(2003).
10. S. Pat, R. Mohammadigharehbagh, S. Özen, Vacuum, **141**(2017).
11. L.X. Su, Y. Zhu, Q.L. Zhang, Applied Surface Science, **274**(2013).
12. J.P. Xiao, T. Frauenhei, J. Phys. Chem. Lett, **3**(2012).
13. H.F. Zhang, S.X. Lu, W.G. Xu, F. Yuan, Surface Science, **625** (2014).
14. M.A. Lahmer, K. Guergouri, Surface Science, **633**(2015).
15. J.P. Perdew, Y. Wang, Phys. Rev. B, **45**(1992) .
16. P. E. Blöchl, Phys. Rev. B, **50**(1993).
17. G. Kresse, J. Hafner, Phys. Rev. B, **47**(1993).
18. G. Kresse, J. Furthermuller,Phys. Rev. B, **54**(1996).
19. X.Q. Dai, H.J. Yan, J.L.Wang, J. Phys: Condens. Matter, **20**(2008).
20. Z. Yang, S.J. Xiong, Surface Science, **605** (2011).
21. X. Y, O.T. Hofmann, R. Patrick, B. Am. Math. Soc, **57**(2012).

Preparation and Performance of Gallium Nitride Powders with Preferred Orientation

Liping Kang[1], *Lingli* Wang[1], *Haiyan* Wang[1*], *Xiaodong* Zhang[1], *Yongqiang* Wang[1]

[1]School of Physics and Electronic Engineering, Zhengzhou University of Light Industry, Zhengzhou City, Henan province, PR China, 450002

Abstract. The paper prepared the III-V semiconductor, hexagonal wurtzite Gallium nitride powders by calcining a gallium oxide in flowing ammonia above 900 °C (1173K). Because of the solid-state reaction process that the gallium oxide transformed to GaN through solid-state gallium oxynitrides (GaO_xN_y) as inter-mediates, the Gallium nitride powders which are agglomerates of tens nanometers flake crystallites retain the rod shape and grain size of raw gallium oxide and have slight (002) plane preferred orientation. The near-edge emission of Gallium nitride at 346 nm has a blue shift of 187 meV attributed to a decrease in disorder of the material that is decided by the (002) plane preferred orientation. The preferred orientation and a blue shift have some kind of reference significance to single crystal growth.

1 Introduction

As a representative of the third generation of semiconductor, Gallium nitride (GaN) has a wide direct gap of 3.39 eV at room temperature, high luminescent efficiency and low temperature-quenching effect, promising application prospects in blue and UV lighting, Ultraviolet Detecting and other areas. Sapphire, Silicon chip and SiC are all the widely used substrates for the thin-film growth of GaN. However, there exists severe mismatch both in lattice parameters and thermal expansion coefficients between GaN and the substrates, which introduce stresses in the first few layers of the grown GaN. Therefore the development of high-purity GaN crystalline for homoepitaxy by wafers is a trend in the future.

The most popular methods to grow single-crystal bulk GaN demand high purity and single phase GaN powder as precursor source, thus one of current significance research topic is the synthesis GaN powder directly. In this research, we took an easy method to synthesize GaN powders, which is based on the reduction reaction of gallium oxides (Ga_2O_3) in ammonia gas.

* Corresponding author: wanghaiyan@zzuli.edu.cn, 2005012@zzuli.edu.cn

2 Experiments

The starting materials include β-Ga_2O_3 (99.999%) and liquid ammonia (NH_3, 99.999%). 0.25 g Ga_2O_3 powders were placed in a quartz tube which was fixed in a furnace. After the air in the tube was expelled by argon (Ar, 99.999%) gas flow for 5 minutes, the furnace was heated to 600 °C at the rate of 10 °C /min with an Ar flow. Then the flow was changed into NH3 mixing in Ar (NH_3/Ar, NH_3:Ar =1:2) and the temperature was increased continuously with a rate of 5 °C/min (K/min) until it reached the reaction temperatures of 850 °C (1123 K), 875 °C (1148 K), 900 °C (1173 K)and 925 °C, (1198 K) respectively. The reaction time under high-temperature condition was 30 minutes. In the cooling process, the NH_3/Ar flow were kept until the temperature reached 600 °C (873 K) and then it was switched to a pure Ar flow.

The as-synthesized GaN samples were characterized by powder X-ray diffraction (XRD) using a Bruker D8 Advanced X-ray diffract-meter with Cu Kα radiation operating at 60 kV and 80 mA. The morphologies of the powders were investigated by JSM-6490LV scanning electron microscope (SEM). The high-resolution transmission electron microscope (HRTEM) lattice images and the corresponding selected-area electron diffraction (SAED) of an individual grain were from JEM-2100 (UHR) HRTEM. The luminous property of GaN was performed by a Horiba FL3 photoluminescence (PL) spectroscopy.

3 Results and discussion

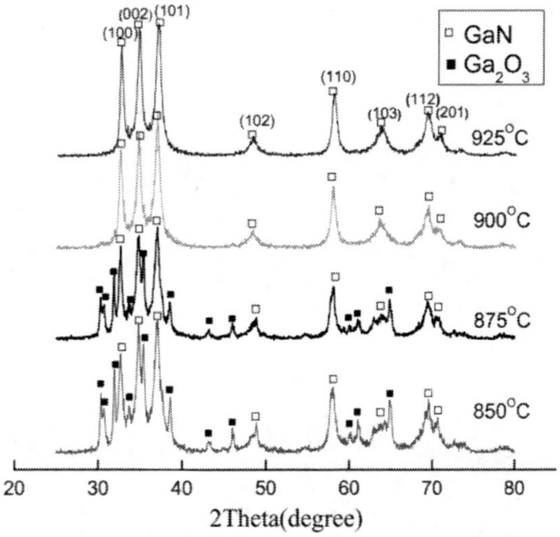

Fig. 1. X-ray diffraction patterns of GaN powders synthesized at different temperatures.

3.1 Structure

Fig. 1 shows the XRD patterns of GaN powder obtained at different temperatures. The XRD patterns show that the samples synthesized at 850 °C and 875 °C are mixtures of hexagonal wurtzite GaN and monoclinic β-Ga_2O_3 (PDF card no. 43-1012 with lattice constants of a = 1.223 nm, b = 0.304 nm, c = 0.58 nm and angles of 90, 103.7, 90 degrees respectively). However, the GaN powders obtained at 900 °C and 925 °C have the typical wurtzite structure. According to PDF card no. 65-3410, the main peaks of the wurtzite structured GaN are indexed to refined lattice constants of a = 0.3173 nm and c = 0.5158 nm.

Even though no other diffraction peak were observed, it was possible that oxygen element existed in GaN samples grown at 900 degree and above because the starting materials were oxides. At the same time, an uncommon phenomenon is seen in the XRD patterns that the diffraction from (002) plans is extra strong compared with the other peaks. This result indicates that the present samples exist preferred (002) crystal plane orientation, like in the thin films [1].

After removing background of curves and fitting peaks processes, the average size of the sub-grains is rough calculated by the full width at half maximum using the Scherrer formula ($t = \lambda/B\cos\theta$), as shown in Table 1; the number in parenthesis is error range. The sizes of GaN crystallite only from GaN diffraction peaks increase with synthesize temperature. The average sizes of powders obtained at 850 °C and 875 °C from the whole spectra are very large because the two samples are mixtures of GaN and large size Ga_2O_3 particles.

Table 1. The crystallite size of samples obtained at different temperatures

Nitride temperature [degree]	850	875	900	925
Average size of powder samples [nm]	310(12)	275(21)	118(2)	126(2)
Average size of GaN crystallite [nm]	69(4)	118(3)	118(2)	126(2)

3.2 Morphology

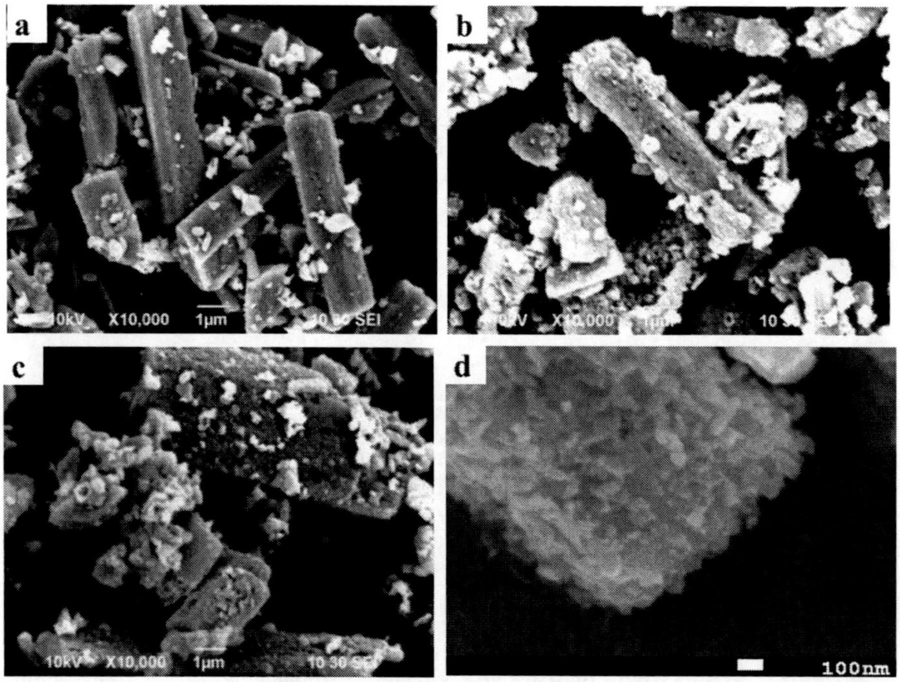

Fig. 2. SEM pictures of Ga_2O_3 (a), GaN samples annealed at 900 °C (b and d), at 875 °C (c).

SEM images of raw material Ga_2O_3 and GaN powders obtained at 900 °C and 875 °C are shown in Fig.2 (a), (b) and (c), respectively. The magnification times are all 10,000. The distribution of the powder size is wide spread from tens of nanometers to several micrometers. GaN samples grown at 900 °C and 875 °C were of rod morphology with a small quantity of GaN nanograins enclosed on its surface. The micrometer rod is consisting of particles about 100 nanometer, as show in Fig.3 (d) GaN at 900 °C details image. The estimated average size is in a good agreement with calculated numbers of XRD. By the comparison of Ga_2O_3 and GaN, we see that both of the shape and size of GaN powders are similar to those of raw material of Ga_2O_3. From this result it can be proved that the GaN powders retain the rod shape and grain size of Ga_2O_3. Thus, a conclusion of the conversion of gallium oxide to GaN don't go through the gaseous species of Ga_2O but solid-state gallium oxynitrides (GaO_xN_y) as inter-mediates is obtained [2].

TEM image, HRTEM image and selected area electron diffraction (SAED) pattern (inset) of the GaN powders at 925 °C are shown in Fig. 3(a) and (b). TEM image shows the presence of irregular agglomerates of tens nanometers of tabular GaN crystallites, being consistent with the calculated results of table 1. The d-spacing between the two adjacent lattice fringes is 0.2586 nm from (002) plane. The calibrated SAED pattern was depicted in the Fig. 2(c). Comparing to those of standard GaN powders and their d-spacing, all the diffraction rings could be attributed to hexagonal wurtzite GaN. The discontinuous diffraction rings demonstrate that the samples exists preferred orientation; this result keeps in line with our XRD measurement.

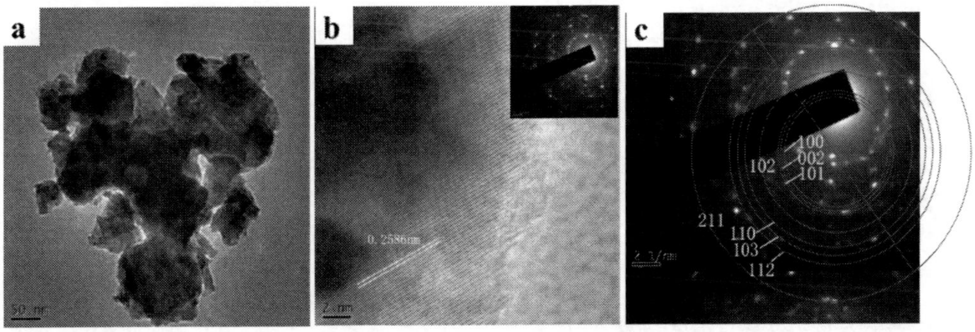

Fig. 3. TEM image (a), HRTEM image (b) and SAED pattern (c) of GaN at 925 °C.

Basing on anionic coordination polyhedron growth units theory model [3], the strong polarity faces grow faster and appear in smaller areas, whereas the weak polarity faces grow slower and appear in large areas [4]. In hexagonal wurtzite GaN, the tetrahedron $[Ga\text{-}N_4]^{8-}$ growing from GaO_4 tetrahedral is considered as the basic unit. Because the vertexes of $[Ga\text{-}N_4]^{8-}$ appear on the (110) faces,the (110) faces show higher surface energy and stronger polarity. On the contrary, the planes of $[Ga\text{-}N_4]^{8-}$ are parallel to (002) faces, so the (002) faces of GaN particles have low surface energy and weak polarity. Therefore (002) faces appear in large areas in the GaN particles growing process. As mentioned earlier, gallium oxide transforms to GaN via gallium oxynitrides (GaO_xN_y) as inter-mediates. As the nitridation proceeds, the oxygen content in the intermediates gradually decreases, leading to complete conversion to GaN and this conversion completed from outside to inside of the rod gradually. The before-mentioned factors result in that there are some GaN plates which are corresponding to the (002) plane attaching on the rod surface. Therefore, the GaN materials obtained have slight (002) plane preferred orientation.

3.3 Luminescence

The luminescent properties of the GaN powders synthesized at 925 °C, 900 °C and 875 °C were examined by photoluminescence (PL) spectroscopy at room temperature, and the PL spectra are shown in Fig. 4a. The exciting source is 294 nm UV.

The dominant emission of the obtained GaN powders is a broad emission from 320 nm to 400 nm centered near 360nm. This broad UV emission band of the GaN powders sample synthesized at 925 °C is fitted with four Gaussians and the peaks are located at 346 nm (3.584 eV), 364 nm (3.407 eV), 365 nm (3.397 eV) and 398 nm (3.116 eV), as shown in the Fig. 4b. The raw date (red hollow square symbols), separated peaks (blue solid line) and the fitted curve (black dotted line) are all provided. The curve fitting result of the GaN powders sample synthesized at 900 °C is similar to that of the GaN powders sample synthesized at 925 °C within the margin of error, and the graph isn't given there.

The emission peak at 346 nm is considered belongs to a blue shift of the near band edge emission at 365 nm (3.39 eV) of the bulk GaN. The blue shift of 19 nm (187 meV) is attributed to a decrease on disorder of the material that is decided by the (002) plane preferred orientation. Considering the separation of the peaks is 174 meV and 10 meV, the 364 nm and 365 nm peaks are possible the electron transition radiation between the shallow donor levels and the valence band [5]. The donor is believed to be a residual oxygen impurity from raw material [5]. The broad violet band centered at 398 nm is relating to defect levels. The well-known yellow band emissions from Carbon impurity deep acceptor defect energy are not observed.

However, the curve of the GaN powders sample synthesized at 875 °C was fitted only with two Gaussians and the peaks were located at 360 nm (3.444 eV), 445 nm (2.787 eV),as shown in the Fig. 4c. The recombination radiation between the shallow donor levels and the valence band will have a red shift and broaden because of the high oxygen content of sample at 875 °C and below [6], then merge with the peak relating to defect levels.

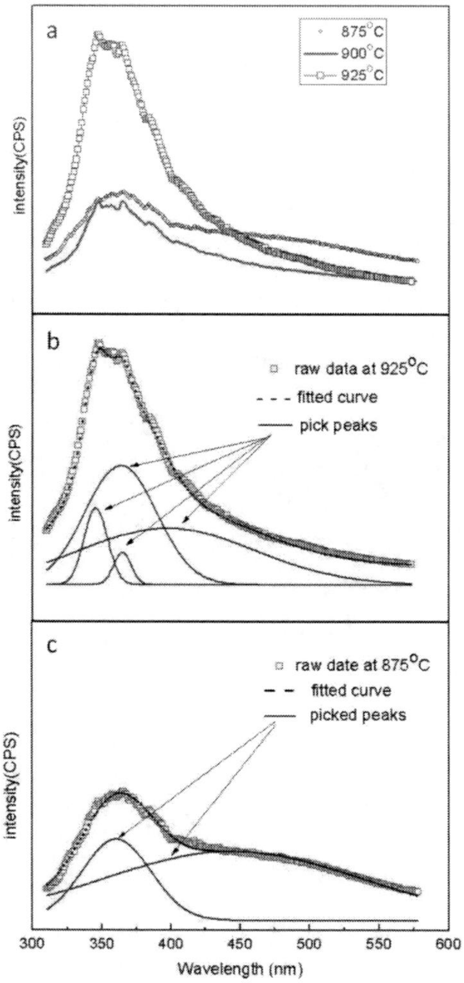

Fig. 4. Room-temperature PL spectra (a) of the powders at 925°C(blue square symbols line), 900°C (red solid line), and 875 °C (black circle symbols), fitted curve graphs of the powders at 925 °C (b), and

4 Conclusions

In summary, GaN powders with hexagonal wurtzite structure are synthesized using a simple and economical method that gallium oxide is nitrided under flow of ammonia at quartz tube furnace. From the XRD and TEM result the present GaN powders have clear preferred orientation of <002>. The similar shape and size of Ga_2O_3 and GaN provide explanation to the growth mechanism of the GaN, i.e. the conversion of gallium oxide to GaN don't go through the gaseous species of Ga_2O but solid-state gallium oxynitrides (GaO_xN_y) as intermediates, and that is the very reason of preferred orientation. The PL spectra show that the near band edge emission of the present GaN is 346 nm (3.584 eV) at room temperature, which exhibits an obvious blue shift of 19 nm (187 meV) because of a decrease in disorder of the material that is decided by the (002) plane preferred orientation. It accounts for the GaN powder with preferred orientation prepared in this method can be used as UV luminescence material.

This research was supported by the Natural Science Foundation of China (No. 51302250), the cutting-edge technology research and development program of Zhengzhou Science and Technology Bureau (No.141PQYJS552), and the Internal Research Fund of Zhengzhou University of light industry (No.2013XJJ010).

References

1. C. B. Han, C. He, X. J. Li, Adv. Mater., **23** 4811-4814 (2011)
2. W. J. Jung, Mater. Lett., **60** 2954-2957 (2006)
3. W. Z. Zhong, H. S. Luo, S. K. Hua, G. S. Xu, J. Synth. Cryst., **33(4)** 475-478 (2004)
4. Z. Y. Zhao, X. G. Li, S. R. Wang, Y. Xiao, J. Synth. Cryst., **43**(7) 1611-1619 (2014)
5. R. Garcia., G. A. Hirata, A. C. Thomas, F. A. Ponce, Opt. Mater., **29** 19-23 (2006)
6. H. S. Guder, B Abay ., H. Efeoglu, Y. K. Yogurtcu, Bulg. J. Phys., **27**(2) 72-75 (2000)

Synthesis and Luminescence Properties of the Binary/Ternary Complex of Samarium Ion with Phenacylmethyl P-Tolyl Sulfoxide and Benzoic Acid

Kui-Suo Yang[1], *Wen-Xian* Li[1*], and *Jin-Rong* Bao[1]

[1]College of Chemistry and Chemical Engineering, Inner Mongolia University, Hohhot 010021, People's Republic of China

Abstract. A new ligand phenacylmethyl p-tolyl sulfoxide (denoted as L_1) has been synthesized by an asymmetric synthesis method. Its novel binary complex $Sm(L_1)_7 \cdot (ClO_4)_3 \cdot 6H_2O$, and ternary complex $Sm(L_1)_6{\cdot}(L_2){\cdot}(ClO_4)_2{\cdot}6H_2O$ (benzoic acid denoted as L_2) have been synthesized. The structure and composition of the samarium complexes characterized by elemental analysis, molar conductivity, IR, UV, and ^{1}HNMR. Additionally, fluorescence spectra measurements shows that the complexes display stronger fluorescence intensity in solid state, indicating the ligands could efficiently transfer the energy to Sm^{3+} ions in the complexes. The ternary complex has stronger fluorescence intensity than the binary complex.

1 Introduction

Rare earth complexes have attracted much attention for a wide variety of applications in the fields of lighting devices, and biomedical analysis [1,2]. There are numerous examples of luminescent rare earth complexes reported in the literature [3,4]. As is known, the intra-configuration 4f→4f transitions in rare earth ions are forbidden, so the luminescence intensity of the RE^{3+} ions are weakening [5]. The organic ligands with great conjugated action the as sensitizers coordinated with the RE^{3+} which can transfer the energy to it effectively and further to increase its luminescence intensity. The organic ligands that exhibited this property were called as "antennas" [6]. In the past several decades, many sulfoxide complexes had been synthesized and intensively studied due to the superb coordination ability of the sulfoxide ligand [7,8]. Europium and terbium complexes are studied a lot. There are few researches about samarium complexes. So it is extremely vital significant for the comprehensive utilization of Sm^{3+} to research the luminescence properties of the samarium complexes.

Thus, we have synthesized a phenacylmethyl p-tolyl sulfoxide (L_1). Furthermore, the L_1 as the first ligand and benzoic acid (L_2) as the second ligand coordinated with samarium ions in the paper. The luminescence properties of as-synthesized binary and ternary

* Corresponding author: nmglwx@163.com

samarium complexes were studied. The luminescence study indicates that ternary samarium complex has higher fluorescence intensity than binary samarium complex.

2 Experimental details

2.1 Materials and physical measurements.

Sm_2O_3 (purity > 99.99 %) was supplied by Beijing Chemical Reagent Company. The Sm(III) perchlorate was prepared from dissolving in 2 mol•L^{-1} $HClO_4$, and then evaporated and dried in vacuum. All other chemicals were analytical grade and were used as received without further purification. Elemental analysis was carried out on a HANAU analyzer. Conductivity measurement was made on a DJS-1 conductivity meter. The infrared spectra were determined on a Nicolet NEXUS-670 FT-IR spectrophotometer. The ultraviolent spectral were recorded on a TU-1901 double beam spectrophotometer. ^{1}HNMR spectra were measured on Bruker AC-500 spectrometer in DMSO-d_6. Fluorescence spectra and fluorescent lifetime were determined on a FLS920 with the slit width of 1nm.

2.2 Preparation of the europium ternary complex.

The L_1 was prepared by oxidation of $C_6H_5COCH_2SC_6H_4CH_3$ according to the literature [9]. The appropriate concentration of L_1 anhydrous ethanol solution was added into samarium perchlorate anhydrous ethanol solution under magnetic stirring for 1.0 h. A white precipitate appeared, the white precipitate was separated by suction filtration, purified by washing with anhydrous ether for several times. Finally, white powder was obtained. The preparation of binary samarium complex was similar to the ternary samarium complex.

3 Results and discussions

3.1 Composition analysis.

The elemental analysis and molar conductivity data of samarium complexes showed in Table 1. The λm value of the binary complex in DMF is in good accordance with the formula as 1:3 electrolyte and the λm value of the ternary complex in DMF is in good accordance with the formula as 1:2 electrolyte [10]. According to the elemental analysis, the compositions of the binary and the ternary complexes are suggested as $Sm(L_1)_7\cdot(ClO_4)_3\cdot 6H_2O$ and $Sm(L_1)_6\cdot(L_2)\cdot(ClO_4)_2\cdot 6H_2O$, respectively.

Table 1. Composition (%) and molar conductivities (S•cm^2•mol^{-1}) of the complexes (25°C)

Complexes	M	Anal. calcd. (found) (%)			λ_m
		C	H	Sm	
$Sm(L_1)_7\cdot(ClO_4)_3\cdot 6H_2O$	2363.33	53.31(53.14)	4.69(4.38)	6.43(6.72)	213
$Sm(L_1)_6\cdot(L_2)\cdot(ClO_4)_2\cdot 6H_2O$	2127.34	54.72(54.77)	4.78(4.45)	7.14(6.91)	183

(The brackets are theoretical values)

3.2 ^{1}HNMR spectra analysis.

The ^{1}HNMR of L_1 shows that there are multiple resonance peaks for two phenyl groups at δ7.375-7.980 ppm, and the integral intensity illustrates it contains nine protons. The proton resonances peaks of one methylene and one methyl group are at δ4.685-4.747 ppm and δ

2.370-2.504 ppm, which have two and three proton respectively (Fig. 1a). The result showed that we have synthesized the L_1 successfully and the purity of the obtained ligand is supernal. The ^{1}HNMR spectra of the samarium complexes are similar as that of L_1, which have shifted at different degree (Fig.1b,c). The reason may be concerned with coordinate effect.

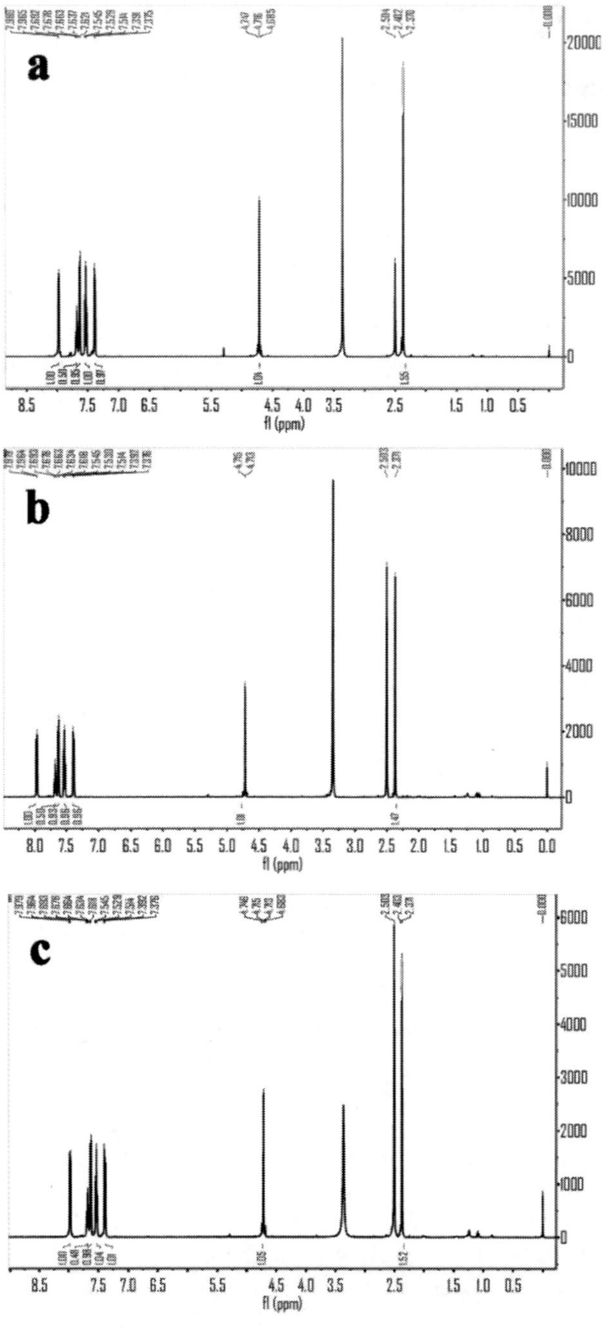

Fig. 1. The ^{1}HNMR spectrum of a. L_1; b. binary samarium complex; c. ternary samarium complex

3.3 Infrared spectra analysis.

Fig. 2a show the IR spectrum of L_1, the stretching vibration absorption of sulflnyl group appeared at 1049 cm^{-1} ($\nu_{S=O}$). It shifts to 982 cm^{-1} and 979 cm^{-1} in IR spectra of binary and ternary samarium complexes (Fig. 2c,d), respectively, which suggests that Sm^{3+} ions are bonded with the oxygen atom of the sulflnyl group. Whereas the carbonyl group stretching vibration absorption in IR spectrum of L_1 (Fig. 2a) appears at 1677 cm^{-1} ($\nu_{C=O}$), no significant shift is found in IR spectra of binary and ternary systems (Fig. 2c,d), which is an evidence that the oxygen atom in carbonyl group is not coordinated with Sm^{3+} ion. The $\nu_{as}(COO^-)$ and $\nu_s(COO^-)$ absorption of L_2 occurred at 1550 cm^{-1} and 1416 cm^{-1}, and the $\Delta\nu$ value is 134 cm^{-1} (Fig. 2b). The $\nu_{as}(COO^-)$ and $\nu_s(COO^-)$ absorption bands appear at 1641 cm^{-1} and 1335 cm^{-1} in the ternary samarium complex and the $\Delta\nu$ value was 306 cm^{-1} larger than that of L_2 (Fig. 2d), which can be deduced that carboxyl group is coordinated with Sm^{3+} ion by monodentate type [11]. In terms of molar conductivities, it can be suggested that none of ClO_4^- is coordinated with Sm^{3+} ions.

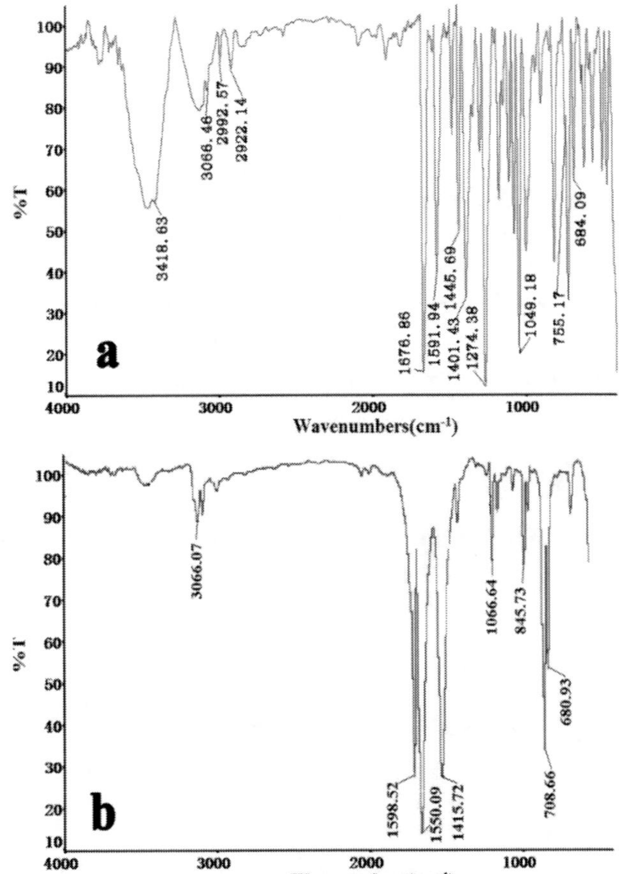

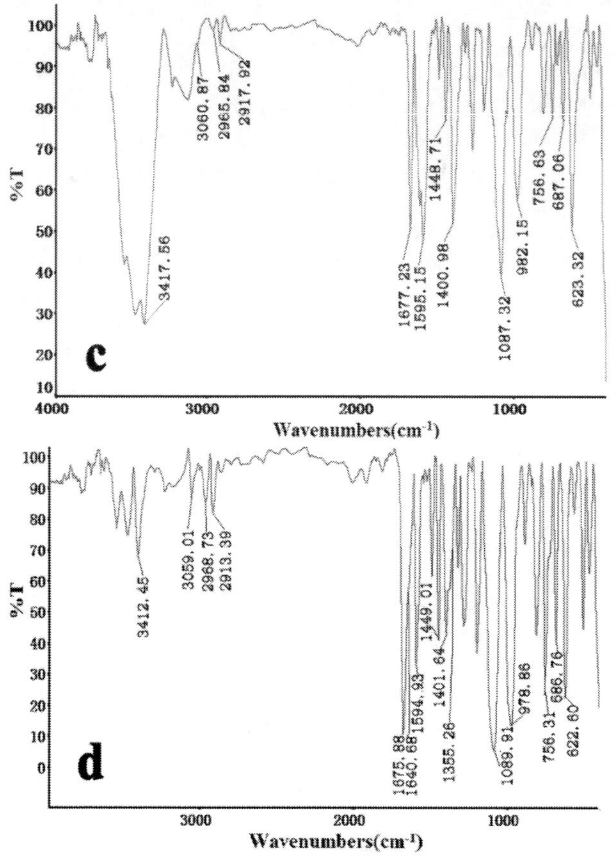

Fig. 2 Infrared absorption spectra of a. L_1; b. L_2; c. binary samarium complex; d. ternary samarium complex

3.4 UV spectra analysis.

The UV spectra of ligands displayed in Fig. 3a and b, the major absorption bands of π-π^* electronic transition in phenyl group are observed at 284 nm and 272 nm, respectively. A strong absorption also observed at 284 nm and 285 nm for the binary and ternary samarium complexes, respectively (Fig. 3c,d), Which is no shift happening compared with the L_1 (Fig. 3b). While the absorption band of binary samarium complex is red shift comparing to the absorption of L_2 absorption band, from 272 nm to 285 nm. It is found that the major absorption bands of the ligands still appeared after the formation of the complexes.

https://doi.org/10.1051/matecconf/201814201010

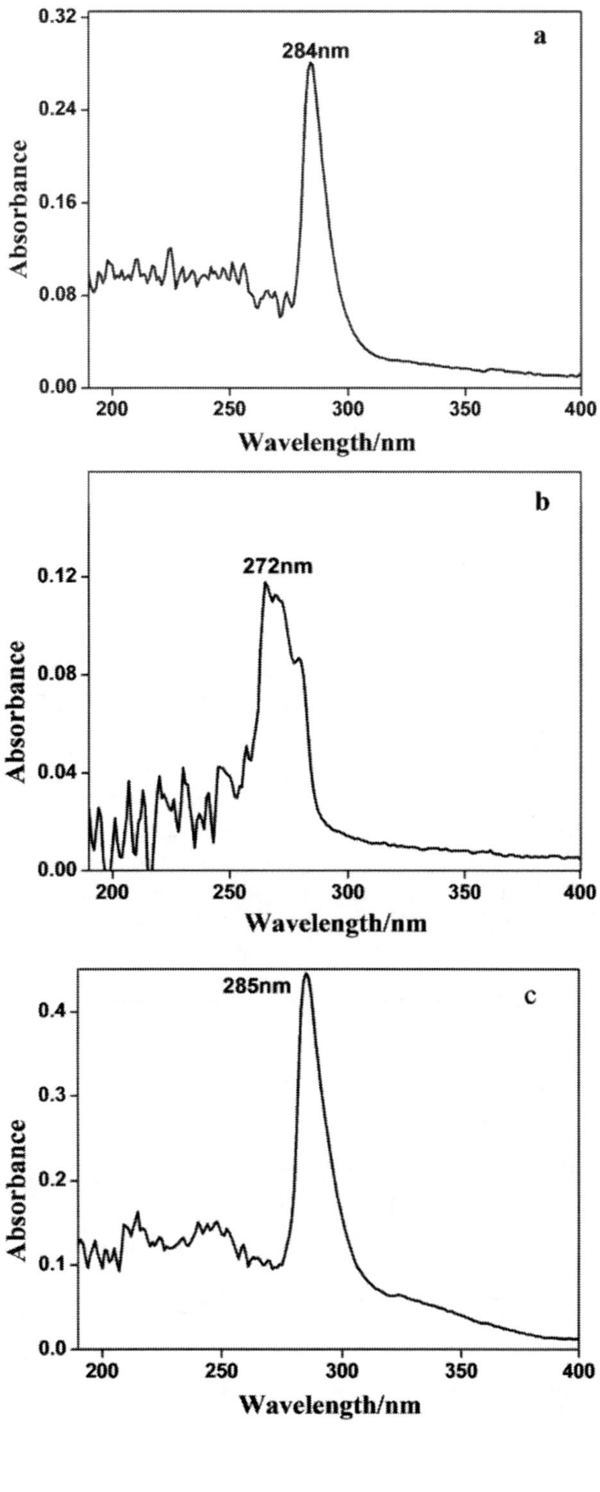

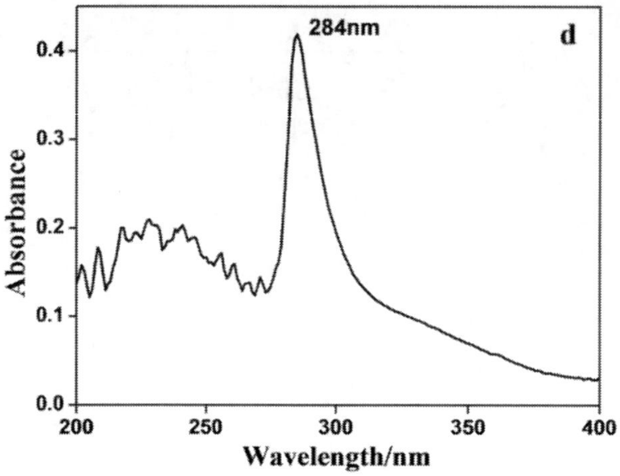

Fig. 3 UV absorption spectra of a. L_1; b. L_2; c. binary samarium complex; d. ternary samarium complex.

3.5 Luminescence properties.

Fig. 4 shows the excitation and emission spectra of binary samarium complex (a) and ternary samarium complex (b). The excitation spectra of the complexes were obtained by monitoring the emission of the complexes at the wavelength of 596 nm. The maximum peak of binary complex was at 317 nm and ternary complex was at 322 nm. The excitation peaks of the two complexes are similar. The emission peaks of the binary and ternary samarium complexes centered at 561, 596 and 643 nm, which are assigned to the transitions of $^4G_{5/2}\rightarrow{}^6H_{5/2}$, $^4G_{5/2}\rightarrow{}^6H_{5/2}$, $^4G_{5/2}\rightarrow{}^6H_{9/2}$, respectively. The binary and ternary samarium complexes display stronger emission intensity, indicating the ligands could efficiently transfer the energy to Sm^{3+} ions in the complexes. The strongest emission intensity of the ternary complex (19680 a.u) is 1.37 times as great as that of the binary complex (14330 a.u.). The result illustrate that the second ligand L_2 can play a synergistic effect for the energy transfer process between the first ligand and Sm^{3+} ion.

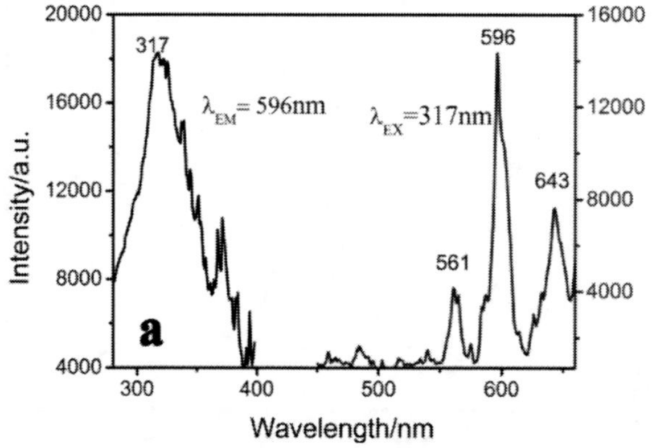

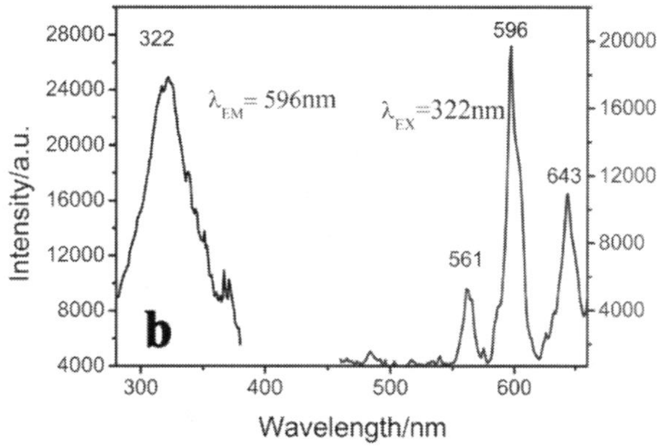

Fig. 4 Fluorescent EX and EM spectrum of a. binary samarium complex; b. ternary samarium complex

Fig. 5 shows the fluorescence decay curves and fit curves of the binary and ternary samarium complexes, respectively. The fluorescence lifetime value of samarium complexes were calculated by the double exponential mode. The lifetime of the ternary samarium complex is determined to be 65.19 μs, which is 1.46 times comparing to that of the binary samarium complex 44.46 μs.

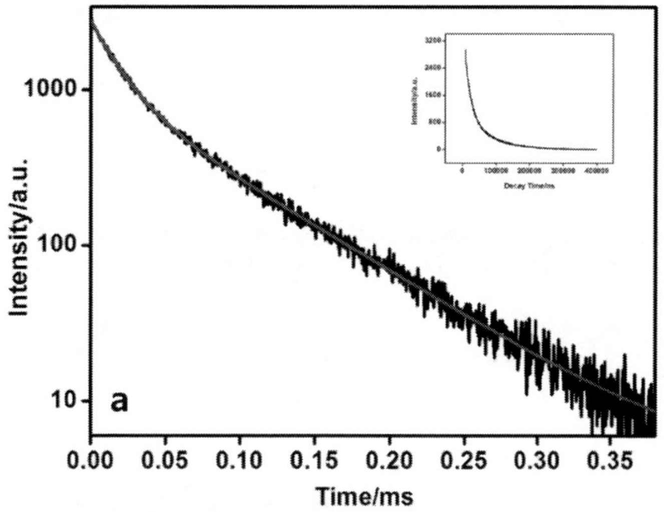

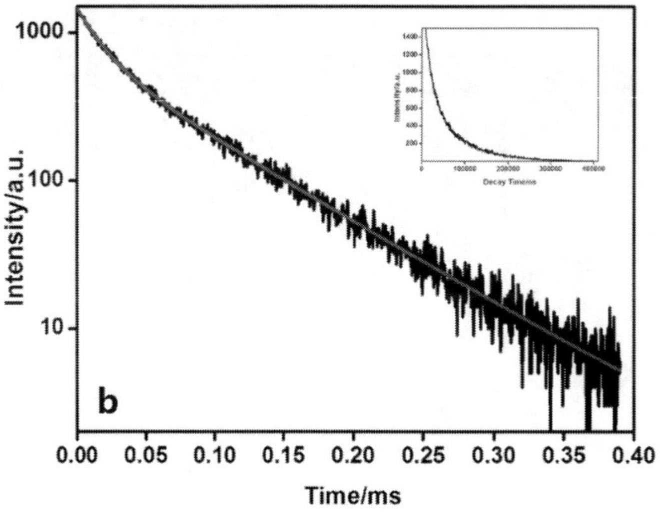

Fig. 5. Linear Fit Curves of a. binary samarium complex; b. ternary samarium complex (Inset: decay curve).

4 Conclusions

In this paper, a new ligand phenacylmethyl p-tolyl sulfoxide has been synthesized by an asymmetric synthesis method. The binary and ternary samarium complexes with phenacylmethyl p-tolyl sulfoxide as first ligand and benzoic acid as second ligand have been synthesized. The samarium complexes display preferable luminescence properties. The emission intensity of the ternary complex is 1.37 times as great as that of the binary complex, and the ternary complex has longer lifetime.

This work is supported by the Major projects of Natural Science Foundations of Inner Mongolia Science Foundation (2015ZD01) and the Natural Science Foundations of Inner Mongolia Science Foundation (2015MS0502).

References

1. A.P. Duarte, M. Gressier, M.J. Menu, et. al, J. Phys. Chem. C. **116** 505-515(2012).
2. L.R. Yang, S. Song, H.M. Zhang, et. al, Synth. Met. **162** 1775-1788(2012).
3. J.H. Xue, X.H. Hua, L.M. Yang, et. al, J. Mol. Struct. **1052** 93-101(2013).
4. A. Gerus, K. Ślepokura, J. Lisowski, Inorg. Chem. **52** 12450-12460(2013).
5. R. Gregory, D.R. Peterman, Coord. Chem. Rev. **174** 283-299(1998).
6. J.M. Lehn, Angew. Chem. Int. Ed. **29** 1304–19(1990).
7. W.X. Li, Y.S. Zheng, X.J. Sun, et. al, J. Fluoresc. **20** 235-241(2010).
8. W.X. Li, Y.S. Zheng, L.H. Saiji, et. al, J. Lumin. **134** 847-852(2013).
9. C. Cardellicchio, O.H. Omar., F. Naso, et. al, Tetrahedron: Asymmetry. **17**:223-229(2006).
10. W.J. Greary, Coord. Chem. Rev. **7** 81-122(1971).
11. G.B. Deacon, R.J. Phillips, Coord. Chem. Rev. **33** 227-250(1980).

Effect of Co-monomer Styrene on Reactive Compatibilization of PE/Wood-flour Composites

Yuewen Li[1*], and *Xinhua* Chen[1]

[1]Hunan Vocational College of Science & Technology, 410004 Changsha, China

Abstract. Compatibilization between polyethylene(PE) and wood-flour was achieved by means of direct reactive extrusion of PE, wood-flour, maleic anhydride(MAH), co-monomer styrene(St) and dicumyl peroxide. Mechanical properties and heat distortion temperature(HDT) of the PE/wood-flour composites were tested, and impact cracking surfaces of the composites were observed via scanning electron microscope. Effect of St on reactive compatibilization of the composites was evaluated. Experimental results indicated that the anchoring strength of interface in the composites was obviously strengthened after St was added. Mechanical properties and HDT of the composites raised with increase of use level of St on the initial stage, and then gradually descended. When use level of St ran up to 1.5%, mechanical properties and HDT of the composites were best. At this moment, the mole ratio of St and MAH was approximately 1:1.

1 Introduction

Wood/plastics composite(WPC) has usually poor compatibility, and compatibilization on WPC can notably raise its property. Dealing with wood-flour via reaction with silane[1], titanate[2], acetic anhydride[3] and benzoyl chloride[4] can reduce polar disparity of wood and plastics, and improve compatibility between them. Macromolecular compatibilizers such as glycidyl methacrylate graft copolymer[5], maleic anhydride graft copolymer[6] and acrylic vinyl monomer copolymer[7] not only contain epoxy group, anhydride and carboxyl which can react with hydroxyl on wood-flour, but also contain macromolecular chain which have favorable compatibility with resin, thus they have commendable compatibilization on WPC.

In the process of melt blending of plastic and wood-flour, direct adding MAH and initiator could improve compatibility of WPC. But percent grafting of MAH on macromolecular chain was low because of complexity of reaction environment, hindering further improvement of WPC property. In this research, it was tried increasing percent grafting of MAH via adding co-monomer styrene(St) to improve compatibilization effect in the process of reaction compatibilization of polyethylene (PE)/wood-flour composites, and influence regular pattern of St dosage on composite property was discussed.

* Corresponding author:lyw1965@126.com

2 The experiment

2.1 Materials

PE, 2911, Wushun Subsidiary Company, Sinopec Group. Wood-flour, particle size 150μm. MAH, A.R., Tianjin Kermel Chemical Reagent Co. Ltd.. St, A.R., Tianjin Damao Chemical Reagent Plant. Dicumyl peroxide(DCP), C.P., Shanghai Lingfeng Chemical Reagent Co. Ltd..

2.2 Preparation method and process parameters

Wood-flour was dried for 12 hour at 100℃ in advance. The raw materials including PE, wood-flour, MAH, St, and DCP were mixed together for 5 min in the SHR high-speed blender. The dosages of components were shown in Table 1. The preliminary mixture was extruded at 180 ℃ via SHJ-20 co-direction twin-screw extruder, then cooled to room temperature and broken up to granules. The final standard test samples were prepared on the GEK80 injection moulding machine. The injection temperature was 180℃, and the injection pressure was 110MPa.

Table 1. Composition of composite samples

Composite number	1#	2#	3#	4#	5#	6#	7#	8#
PE(wt%)	70	68.74	68.39	68.04	67.69	67.34	66.99	66.64
Wood-flour(wt%)	30	29.46	29.31	29.16	29.01	28.86	28.71	28.56
MAH(wt%)	0	1.5	1.5	1.5	1.5	1.5	1.5	1.5
DCP(wt%)	0	0.3	0.3	0.3	0.3	0.3	0.3	0.3
St(wt%)	0	0	0.5	1.0	1.5	2.0	2.5	3.0

2.3 Morphology observation and properties testing

Impact cracking surfaces of samples were observed with QUANTA-200 SEM after sputter-coated with platinum. HDT was tested according to ISO75-2:2003(A) on RV-300 thermal deformation Vicat softing point temperature tester. Tensile and flexuous tests were respectively performed according to GB/T 16421-1996 and GB/T 9341-2000 on WDW3020 micro-control electron universal test machine. Impact tests were performed according to GB/T 1043-1993 on XJC-25ZD electron impact tester.

3 Results and discussions

3.1 Reactive compatibilization mechanism of the composite

In melting extrusion process of PE, wood-flour, MAH and DCP, DCP decomposed to produce primary free radical, and it captured hydrogen atom on PE molecule chain, resulting of creation of PE macromolecule free radical, then it reacted with MAH to create MAH grafted PE(PE-g-MAH). PE long chain of PE-g-MAH could integrate into PE matrix, and MAH radical group could react with hydroxyl on wood-flour, thus the wood-flour and plastic were bonded together. In melting graft process of MAH with PE, percent grafting of MAH was usually low, and cograft of proper second monomer could effectively raise its percent grafting[8,9]. Co-monomer is usually electron rich monomer, such as St. St could more quickly react with PE macromolecule free radical than MAH, and generated St radical could exist longer, and it could more rapidly react with MAH than PE macromolecule free radical, resulting of increase of percent grafting of MAH. In other research, it was considered that MAH could form electron transfer complex with St[10], then it could more rapidly be grafted onto PE macromolecule chain. Addition of St speeded up grafting reaction between macromolecule free radical and MAH, and reduced cross-linking opportunity between macromolecule free radicals, and prevented degradation of processing properties of composites. Compatibilization reactions of MAH and St on the composites are shown in Fig.1.

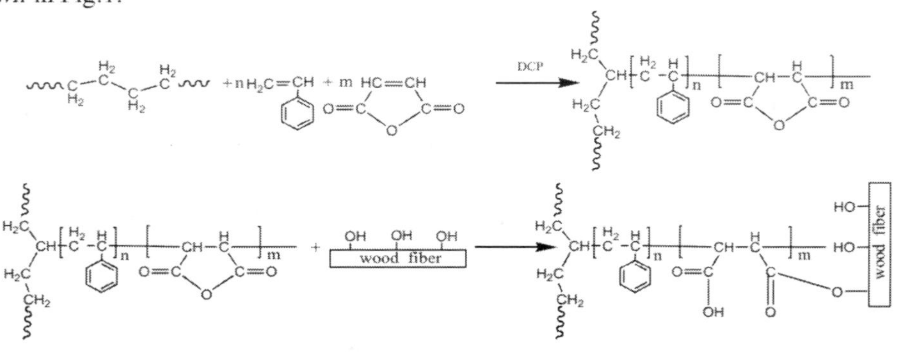

Fig. 1. Compatibilization reactions of MAH and St on the composites

3.2 Morphology of impact cracking surfaces of the composites

SEM micrographs of impact cracking surfaces of the composites are shown in Fig.2. Fig.2(a) is the SEM micrograph of impact cracking surface of 1# WPC, which wasn't compatibilized. It can be observed that there is obvious gap between wood-flours and matrix, which indicates weak interfacial adhesion between PE and wood-flours in the non-compatibilized composites. Fig.2(b) and (c) are respectively the SEM micrographs of impact cracking surface of 2#,and 5# composites. In melting extrusion process of 2# composites, MAH and DCP were added; it can be observed that combination of PE and wood-flours is compact. But peeling phenomenon between wood-flour and PE appeared under the action of impact force, which indicates that matrix stress didn't completely be transmitted to wood-flour. In melting extrusion process of 5# composites, MAH, DCP and co-monomer St were added, it can be observed that interface between PE and wood-flour is obscure, and peeling phenomenon didn't appear under the action of impact force, which indicates that addition of St improved compatibilization effect, and heightened bonding strength of the composites.

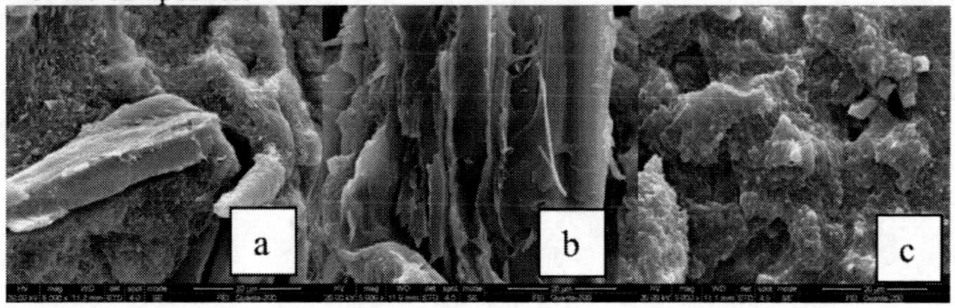

Fig. 2. SEM micrographs of impact cracking surfaces of the composite samples: (a)1#, (b)2#, (c)5#

3.3 Effect of reactive compatibilization on mechanical property and HDT of the composites

Mechanical property and HDT of 1# and 2# composites are shown in Table 2. 1# composites didn ' t carry through compatibilization treatment, and 2# composites carried through reactive compatibilization via MAH and DCP. It can be found that tensile strength, flexural strength, notched impact strength, elongation at break and HDT increased obviously, which indicates that direct reactive compatibilization can effectively improve property of the composites. After direct reactive compatibilization, compatibility of the composites increased, and combination between PE and wood-flour become close together, therefore matrix stress could effectively be transmitted to wood-flour, resulting of increase of tensile strength, flexural strength, notched impact strength and elongation at break. When heated, matrix stress of the composites after compatibilization could be transmitted to wood-flour as before, and wood-flour resistance to thermal deformation excels obviously PE, thus HDT of the composites heightened evidently.

Table 2. Comparison of mechanical property and HDT of the composites before and after reactive compatibilization

WPC number	Tensile strength (MPa)	Flexural strength(MPa)	Elongation at break(%)	Notched impact strength(KJ/m^2)	HDT(℃)
1#	19.2	23.7	8.4	3.5	70.4
2#	32.8	29.7	11.5	5.1	78.2

3.4 Effect of St dosage on mechanical property of the composites

Influence law of St dosage on mechanical property of the composites is shown in Fig.3-Fig.6. When St dosage was less, tensile strength, flexural strength, notched impact strength and elongation at break of the composites increased with the increase of St dosage. Mechanical properties of the composites were the best at 1.5% dosage of St. After that, with the increase of St dosage, mechanical properties of the composites decreased gradually. When MAH grafted with PE alone, its grafting points were less, and graft chain length was shorter, resulting of low grafting rate of MAH. After adding co-monomer St, MAH could be grafted onto PE molecular chain faster and more with St as mediator, and grafting rate increased, resulting that bonding strength of the composites increased, and the mechanical properties were also improved. When the St dosage reached 1.5%, the molar ratio of St to MAH was about 1:1. This moment, whether St and MAH reacted with the macromolecular free radicals one after another, or St and MAH in the form of electron transfer complexes reacted with the free radical, St could give full play to the role of bridge between MAH and PE, and the grafting rate of MAH was highest, and bonding strength of the composites was firmest, and the mechanical properties were also best. When the amount of St continued to increase, there were some excess St molecules. Because of the existence of a competitive relationship between excess St molecules, MAH and electron transfer complexes, the chance of MAH grafting with PE was reduced, resulting that the grafting rate of MAH and mechanical properties of composites decreased gradually.

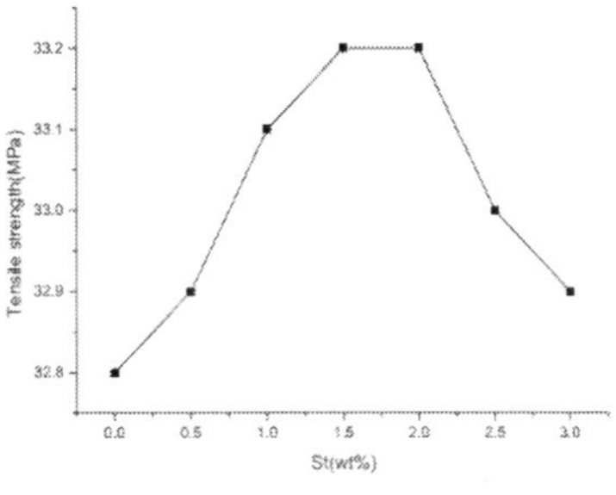

Fig. 3. Effect of St dosage on tensile strength

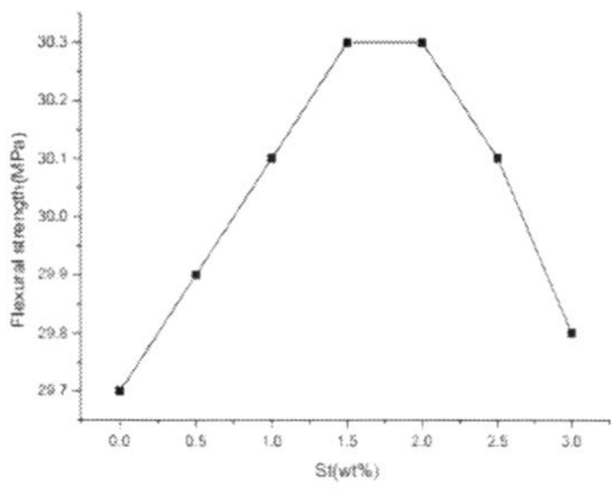

Fig. 4. Effect of St dosage on flexural strength

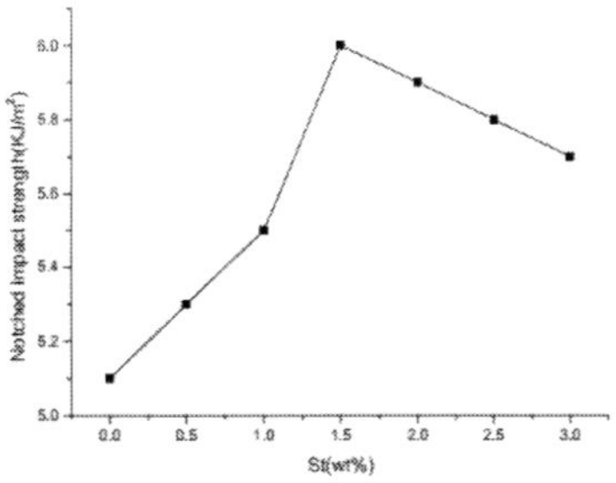

Fig. 5. Effect of St dosage on notched impact strength

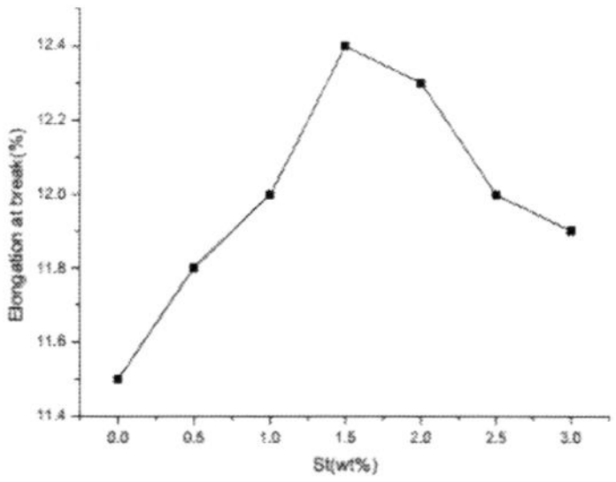

Fig. 6. Effect of St dosage on elongation at break

3.5 Effect of St dosage on HDT of the composites

Relationship between HDT of the composites and St dosage is shown in Fig.7. The variation of HDT with the St dosage was consistent with the mechanical properties of the composites. HDT of the composites reflected resistance to bending deformation under heating conditions. Resistance of PE matrix of the composites to bending deformation under heating conditions descended, but resistance of wood-flours was less affected by temperature. Grafting ratio of MAH was greater, combination of the composites was more compact, and stress born by PE matrix could be more transferred to wood-flours, and HDT of the composites was higher.

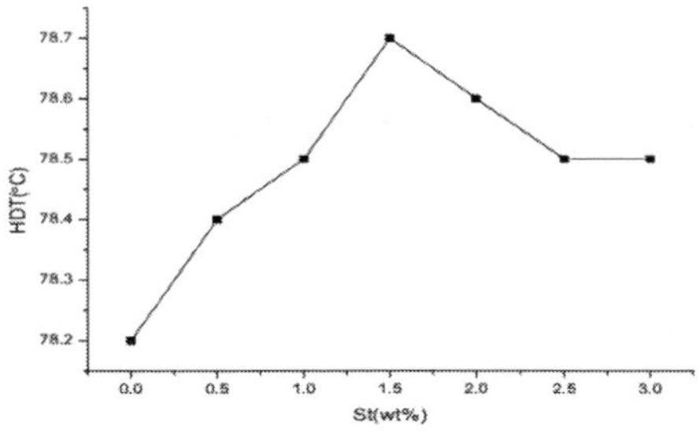

Fig. 7. Effect of St dosage on HDT of the composites

4 Conclusions

(1) SEM images showed that combination of the composites was more compact after adding the co-monomer St in the process of reactive compatibilization.

(2) Mechanical properties and HDT of the composites were significantly improved after reactive compatibilization.
(3) In the process of reactive compatibilization, mechanical properties and HDT of the composites increased with the increase of St dosage at initial stage, then gradually decreased. When use level of St ran up to 1.5%, mechanical properties and HDT of the composites were best. At this moment, the molar ratio of St to MAH was about 1:1.

References

1. M. B. N. Sonia, S. C. Graziela, M. L. R. Simone, Polymer Testing, **26,** 619(2007)
2. L. L. Yang, H. Li, Z. Y. Zhong, Plastics Science and Technology, **38**, 36(2010)
3. T. F. Qin, H. P. Yan, China Wood Industry, **13**,17(1999)
4. R. C. Mankikadan, S. M. Nair, T. Sabu, Jorunal of Applied Polymer Science, 1483(1996)
5. C. Xiong, R. R. Qi, Engineering Plastics Application, **36**, 22(2008)
6. T. Liu, S. R. Lu, Y. L. Wang, C. X. Zhang, Z. Y. Huang, Plastics Science and Technology, **38**, 58(2010)
7. R. Y. Wang, C. Y. Wan, Y. Zhang, Y. X. Zhang, China Plastics, **19**, 63(2005)
8. X. M. Xie, N. H. Chen, S. Li, Acta Polymerica Sinica, 351(1999)
9. B. D. Zhu, J. Wang, H. L. Li, W. C. Dong, C. Dong, China Plastics, **26**, 50(2012)
10. Y. Li, X. M. Xie, B. H. Guo, Polymer, **42**, 3419(2001)

Analysis on Interfacial Performance of CFRP-Concrete with Different Thickness of Adhesive Layer and CFRP Plate

Qingyong Guo[1] *, *Huijian* Zhao[1], *Lei* Chen[1], *Jize* Mao[1], and *Wensong* Zhang[2]

[1]College of Aerospace and Civil Engineering, Harbin Engineering University,Harbin 150001, China
[2]General Research Institute for Nonferrous Metals, Beijing 100088, China

Abstract. The bond behavior of CFRP-concrete interface is the analysis foundation for concrete structures with external strengthening CFRP. In the paper, the influences of the thickness of CFRP plate and adhesive layer on interfacial adhesive properties are investigated through the finite element program. The influence rules of the thickness on the interfacial ultimate bearing capacity and the effective bond length are performed.The results show that the thickness of adhesive layer and CFRP plate has a significant effect on the interfacial performance of CFRP-concrete.

1 Introduction

In recent years, the strengthening of building structure has become the focus of attention. At present, the commonly used reinforcement methods are increasing the old section method, cladding steel method, carbon fiber (Carbon Fiber, Reinforced, Polymer, referred to as CFRP) reinforcement method. Carbon fiber reinforcement has attracted wide attention due to its superior tensile and durability properties. Therefore, the study on interfacial properties of CFRP-concrete plays an important role in analysis on the performance of CFRP reinforced concrete structures.

At present, researchers in China and abroad have made a great deal of studies on the interfacial performance of CFRP and concrete [1]. It is shown that the thickness of CFRP plate and interface adhesive layer have great influences on the interfacial bond behavior of CFRP-concrete. Otherwise, there arefew researches focus on the effects of the thickness of interface adhesive layer on bonding properties for CFRP- concrete. Peng Hui etc. concluded that the increase of the adhesive layer thickness reduced the fatigue damage, and the ultimate bearing capacity also increased [2]. Ma Tao etc. carried out single shear test finding that the initial peel load increases with the thicker adhesive layer[3].It is found that the layer thickness is one of the most important parameters for interfacial adhesive properties of CFRP and concrete,and there are differences among existing conclusions. As a result, there is need to do further study on it. The paper studies the influences of the thickness of CFRP plate and adhesive layer on interfacial adhesive properties by dint of the numerical analysis. The influence rules of the thickness on the interfacial ultimate bearing

* Corresponding author: qingyong622@126.com

capacity and the effective bond length are performed. Some conclusions have been drawn finally.

2 Constitutive model of materials

2.1 Plastic damage constitutive model of concrete

The plastic damage model of ABAQUS is used for concrete. The model is based on the Lubliner[4] and Lee and Fenves[5] models. The model expresses the inelastic behavior of concrete materials using an isotropic damage elasticity combined with an isotropic tensile and compressive plastic model. The model considers the degradation of elastic tensile and compression stiffness strain caused by plastic strain. It can simulate the mechanical properties of concrete, such as crushing and tension fracture, and is a plastic model based on continuous medium[6].

2.2 Constitutive model of interface element

Based on the analysis of nonlinear fracture mechanics, it can be seen that the bearing capacity of CFRP- concrete interface is related to the fracture energy of interface, and it is independent of the shape of bond slip curve[7]. Because the bilinear model is simple in form, it is easy to be used in finite element simulation. The bilinear model has three important parameters, namely, the initial shear stiffness of interface(K_0), the maximum shear stress at interface(τ_{max}) and the interfacial fracture energy(G_f).

Interfacial initial shear stiffness can be expressed as

$$K_0 = \frac{\tau_{max}}{s_0} \tag{1}$$

Where, s_0 is the corresponding relative slip of the maximum shear stress.

Interfacial fracture energy can be expressed as

$$G_f = \frac{1}{2}\tau_{max} s_f \tag{2}$$

Where, s_f is the maximum slip of the interface.

2.3 The interface bond slip model

Many researchers obtained different interface bond slip models through experimental study. Here Ueda et al. modelis adopted for simulation which considers more factors. It is expressed as follows,

$$\tau = 2BG_f\left(e^{-Bs} - e^{-2Bs}\right) \tag{3}$$

$$\tau_{max} = 0.5BG_f \tag{4}$$

$$s_0 = \frac{0.693}{B} \tag{5}$$

$$B = 6.846\left(E_f t_f\right)^{0.108}\left(G_a / t_a / 1000\right)^{0.833} \quad (6)$$

$$G_f = 0.446\left(E_f t_f\right)^{0.023}\left(G_a / t_a / 1000\right)^{-0.352} f_c^{\,0.236} \quad (7)$$

Where, E_f is the elastic modulus of the FRP plate, t_f is the thickness of the CFRP plate, and G_a is the shear modulus of the adhesive layer, and t_a is the thickness of the adhesive layer, s is the local slip, f_c is the compressive strength of concrete.

3 Finite element model

At present, the methods to analyze the interface performance of CFRP-concrete by using finite element software can mainly be classified into two kinds. One is to divide concrete and CFRP into the same size units in the process of building the finite element model. Common nodes are used to establish a connection between the concrete and CFRP board, where the appropriate constitutive model of concrete is selected. The process of CFRP-concrete interfacial debonding is reflected by the cracking ofconcrete on the interface.

The second method is to transform the nonlinear problem of concrete cracking at the interface into the failure of the interface element. It not only saves the computing cost, but also will consider the effect of adhesive thickness on the interfacial properties of CFRP-concrete. Therefore, the paper introduces the method of introducing special interface unit between CFRP plate and concrete. Fig. 1 is the geometrical size and load position of the model.

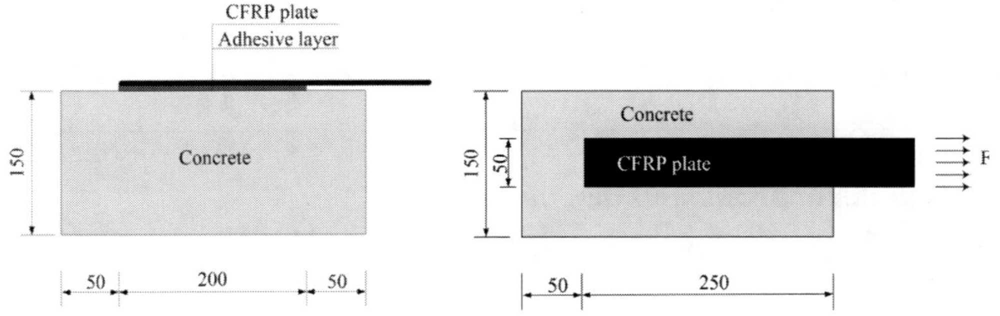

Fig. 1. FRP-to-concrete bonded joints model

Table 1. Material performance index [8]

Material	Project	Parameter
CFRP plate	Elastic modulus	165000 MPa
	Tensile strength	2461 MPa
	Elongation	1.71%
Epoxy resin	Elastic modulus	2627 MPa
	Tensile strength	31.7 MPa
	Elongation	1.5%
Concrete	Strength	38.5MPa

The width, height and length of the concrete specimen are 150mm, 150mm and 300mm separately. The width of CFRP plate is 50mm, the bond length is 200mm, and the loading end has 50mm from nonbonding zone. In the paper, the concrete model is made of solid element, and the element type is chosen as the C3D8R unit with 3D eight nodes which are reduced integration element. In the entire model from stress to failure, CFRP are in the small strain condition. For convergence and convenience of the model, the CFRP plate can regard as a linear elastic material. It does not distinguish between fiber and resin, and uses the equivalent elastic modulus for modeling. The adhesive layer takes COH3D8 unit as cohesive element with 3D eight nodes. The material parameters which are used for the model are shown in table 1. Parameters of specimen and ultimate bearing capacity calculated are shown in table 2.

Table 2. Parameters of specimens and numerical ultimate bearing capacity

Specimen number	CFRP plate thickness (mm)	Adhesive layer thickness (mm)	Interface initial stiffness (MPa/mm)	Maximum shear stress (MPa)	Interfacial failure energy (J)	Numerical ultimate bearing capacity(kN)
F1-1	1	1	119.71	6.69	1080	33.55
F1-2	1	2	48.15	4.80	1378	34.22
F1-3	1	3	28.26	3.95	1590	32.67
F1-4	1	4	19.37	3.44	1759	29.88
F1.4-1	1.2	1	129.74	6.99	1088	39.11
F1.4-2	1.2	2	52.18	5.01	1389	39.09
F1.4-3	1.2	3	30.63	4.12	1602	35.98
F1.4-4	1.2	4	20.99	3.59	1773	32.45
F2-1	2	1	141.28	7.33	1097	46.66
F2-2	2	2	56.83	5.25	1400	44.50
F2-3	2	3	33.35	4.32	1615	39.43
F2-4	2	4	22.86	3.76	1787	34.80

4 Results analysis and discussion

4.1 Analysis of ultimate bearing capacity

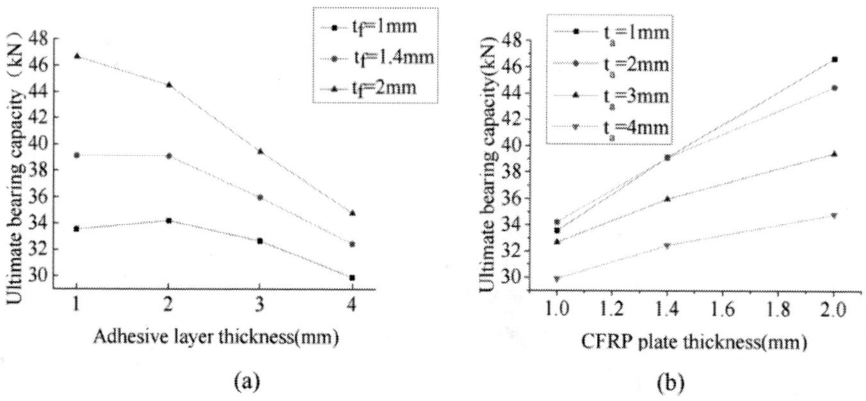

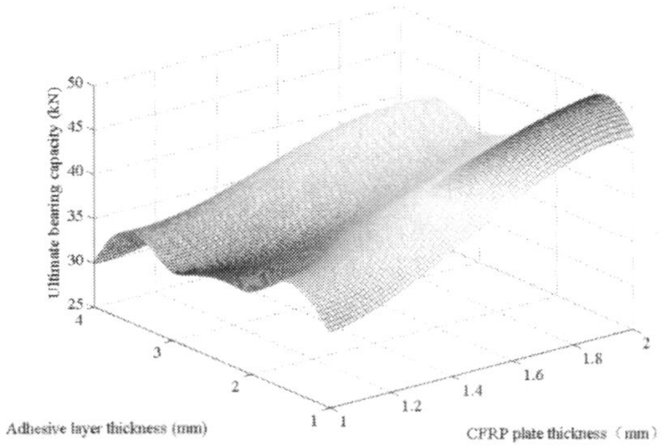

(c)

Fig. 2. Ultimate bearing capacity of interface with different thickness of adhesive layer and CFRP

In Fig. 2 and table 2, it is shown that the thickness of the adhesive layer and CFRP plate have a great influence on the interfacial ultimate bearing capacity. The ultimate bearing capacity of the interface increases generally as the thickness of the adhesive layer decreases and as the thickness of the CFRP plate increases.Fig. 2(a) shows that the ultimate bearing capacity of 1mm CFRP plate increases firstly and then decreases with the increase of the adhesive layer thickness, and reaches the maximum around 2.0mm. However, it is almost flat for 1.4mm CFRP plate and goes down for 2.0mm CFRP plate.

Fig. 2 (b) shows that the ultimate bearing capacity increases with the increase of the plate thickness.When the adhesive layer thickness is 1mm and 2mm, the interfacial ultimate bearing capacity increases with the increase of the plate thickness, which is obviously greater than that of the adhesive layer at 3mm and 4mm. Its maximum value can be obtained when the plate thickness is around 2.0mm and adhesive layer thickness is around 1.5mm as shown in Fig. 2 (c).

4.2 Distribution of strain in CFRP plates and adhesive layers

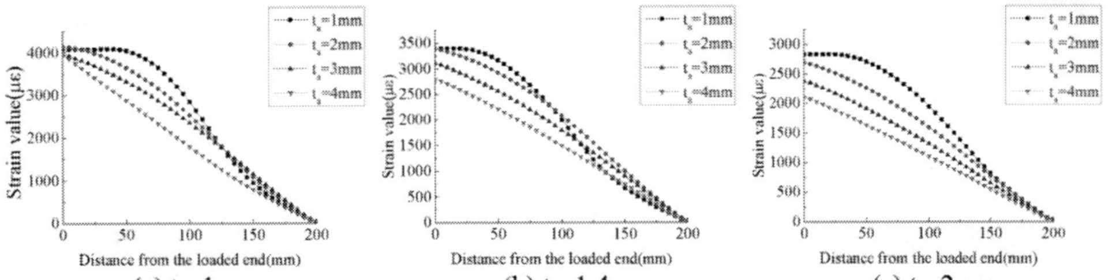

(a) t_a=1mm (b) t_a=1.4mm (c) t_a=2mm

Fig.3. Strain distribution in CFRP plates

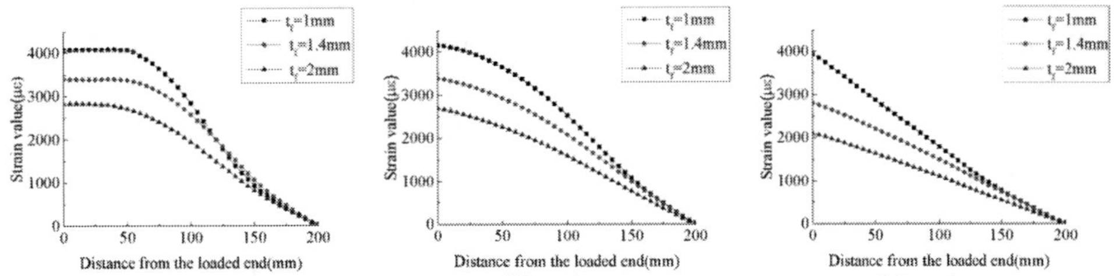

(a)t_f=1mm (b) t_f=2mm (c) t_f=4mm

Fig.4. Strain distribution in adhesive layers

Fig. 3 exhibits the variation of strain values on the CFRP plate with ultimate bearing capacity. It is demonstrates that the maximum strain in the plates decrease as the thickness of plate increases.Fig. 3(a) shows that the maximum strain of the four specimens has little difference when the thickness of the CFRP plate is 1mm. The flat segments on the strain curve mean that the bond length of the two specimens is greater than the effective bond length when tf=1mm and tf=2mm. It explains why the trend of the interfacial ultimate bearing capacity goes flatly when tf≤2mm. In Fig.3(c), the maximum strain of the four curves decreases with the increase of the adhesive layer, and only the curve with the adhesive layer of 1mm has a horizontal section. The above results show that the increase of the adhesive layer thickness leads to the increase of the interfacial effective bond length.

Fig. 4 gives the strain distribution of the CFRP plates with different thickness. The lengths of flat segment are 50mm, 45mm and 30mm for ta=1mm, ta=1.2mm and ta=1.4mm respectively shown in Fig. 4(a). It indicates that the effective bond lengths of the three specimens are 150mm, 155mm and 170mm respectively when tf=1mm. Fig. 4(b) and Fig. 4(c) show that lengths of the flat segments become short and disappear finally. It is found that the effective bond length increases with the increase of plate thickness. This corresponds to the trend of the curve in Fig. 2.

5 Summary

In the paper, the numerical analysis had been carried out on interfacial performance of CFRP-concrete with the CFRP plate thickness 1mm, 1.4mm, 2mm and adhesive thickness 1mm, 2mm, 3mm and 4mm, the following conclusions are obtained:

1)The thicknesses of adhesive layer and CFRP plate have a significant effect on the interfacialperformance of CFRP-concrete.

2)The ultimate bearing capacity of the interface increases generally as the thickness of the adhesive layer decreases and as the thickness of the CFRP plate increases. Its maximum value can be obtained when the plate thickness is around 2.0mm and adhesive layer thickness is around 1.5mm

3)The increase of the thickness of adhesive layer and CFRP plate leads to the increase of the interfacial effective bond length evidently.

This research was financially supported by Natural Science Foundation of Heilongjiang Province of China(NO:E201415) and the Fundamental Research Funds for the Central Universities of China (No.GK2020260143).

References

1. Guo Zhanggen, Sun Weimin, Zhen Zhen. Journal of Nanjing University of Technology (Family Science Edition). **28**,37 (2006)
2. Peng Hui, Gao Yong, Xie Chao, et al. Experimental mechanics. **29**,489 (2014)
3. Ma Tao, Pan Jinlong, Wei Hongli. Building Structure. **19**,15 (2013)
4. Lubliner J, Oliver J, Oller S, et al. International Journal of Solids and Structures. **25**,299 (1989)
5. Lee J, Fenves G L. Journal of Engineering Mechanics. **124**,892 (1998)
6. *ABAQUS analysis user's manual* v6.14 (ABAQUSInc, 2014)
7. Täljsten B. *Springer* (2014)
8. Peng Hui, Gao Yong, Xie Chao, et al. Experimental mechanics. **29**,489 (2014)

Design Method for Proportion of Cement-Foamed Asphalt Cold Recycled Mixture

Junxiao Li [1*], *Wei* Fu [2], and *Hechao* Zang [1]

[1]Shandong Provincial Key Laboratory of Ocean Environmental Monitoring Technology, Institute of Oceanographic Instrumentation, Shandong Academy of Sciences, Qingdao 266001, China
[2]Navy Qingdao First Sanatorium, Qingdao 266001, China

Abstract. Through foaming experiment of Zhongtai AH-70 asphalt, the best foaming temperature water consumption and influence factors of foamed asphalt's foaming features are determined; By designing the proportion of foamed asphalt cold in-place recycled mixture combined with the water stability experiment, for this mixture the best foamed asphalt addition is 3%, and proportion of the mixture is RAP: fine aggregate: cement=75:23:2. Using SEM technology, the mechanism of increasing on the intensity of foamed asphalt mixture resulted by the addition of cement was analysed. This research provides reference for cement admixture's formulation in the designing of foamed asphalt cold in-place recycled mixture.

1 Introduction

In China, the total length of the national road was 469.63 million kilometres by the end of 2016. So there will be a large amount of recycled asphalt mixture in the road pavement renovation, renovation and conservation every year. If calculating by 12% of highway pavement need repair, there will be 2.2 million tons recycled asphalt mixture every year[1]. Asphalt cold recycling can be a good solution, which can make the old pavement materials reused and save 40%-50% of the total investment[2]. Foamed asphalt is the bituminous material that contains a large amount of evenly dispersed bubbles after expansion. In a special foaming device, the expansion occurs when the heat asphalt comes into contact with the water. Now, foamed asphalt is used as a regenerative agent for recycled RAP. When foamed asphalt comes into contact with the aggregate, it suddenly becomes millions of tiny particles and spreads on the surface of the fine aggregate especially particle size less than 0.075 mm. This kind of the fine aggregate with foamed asphalt can be filled in the gap between the coarse aggregate and after mixing and compacting, the asphalt rubber paste is formed that makes the mixture a certain cohesion and strength[3,4]. In recent years, the cold regeneration technique of recycled waste road materials by foamed asphalt has been famous for its remarkable advantages in energy saving, environmental protection and economy[5,6,7]. Before using foamed asphalt cold regeneration technique to renovate road pavement, the proportion of mixture should be designed by laboratory research. Through foaming

* Corresponding author: caike0501@163.com

experiment of Zhongtai AH-70 asphalt, the best foaming temperature, water consumption and influence factors of foamed asphalt's foaming features are determined; By designing the proportion of foamed asphalt cold in-place recycled mixture combined with the water stability experiment, for this mixture the best foamed asphalt addition cement addition and proportion are determined.

2 Experiment

2.1 Raw materials

Foamed asphalt is Zhongtai AH-70 asphalt. Table 1 is the technical indicators of Zhongtai AH-70 asphalt. The recycled asphalt pavement (RAP) is from highway milling material in Wuhan, China. Fig. 1 is the particle size distribution of stabilized soil and RAP. Stone dust is 0 ~ 3mm limestone. The mine powder is limestone powder. Cement is P.O. 425 ordinary Portland cement from Huaxin cement factory, table 2 is its chemical composition. Water is just general water.

Table 1. Technical indicators of Zhongtai AH-70 asphalt.

Technical indicators	Test results	Specification requirements
Penetration (25℃,100g,5s)/0.1mm	65	60~80
Ductility (5cm/min,15℃)/cm	>100	>100
Softening point (ball and ring method)/℃	48	>47

Table 2. Chemical composition of cement.

Chemical composition	SiO_2	Al_2O_3	Fe_2O_3	CaO	MgO	Na_2O	SO_3	Ignition loss
Content /%	21.73	4.25	2.67	63.59	2.21	0.55	2.39	1.75

2.2 Preparation of foamed asphalt

Foaming experiment of Zhongtai AH-70 asphalt choosing 150℃, 160℃ and 170℃ three kinds of foaming temperature, the water consumption at each temperature was 1.5%, 2.0%, 2.5% and 3% respectively. The water consumption used is based on asphalt content. Every experiment were done three times, take the average result to measure foamed asphalt's expansion ratio and half-life.

2.3 Specimen preparation and test methods

By comparing the particle size distribution of RAP with engineering design level gradation of foamed asphalt cold in-place recycled mixture, the mixture gradation is determined. Foamed asphalt cold in-place recycled mixtures have five different cement compositions of 0%, 1%, 1.5%, 2%, 3%. The optimized amount of water used and the maximum dry density were determined by the compaction test, and the mixing water consumption is calculated. The impaction test was referring to the method in *Highway Engineering stable inorganic binder materials testing procedures* (JTG E 51-2009) the preparation of mixture standard

Marshall Specimens was refer to the methods in *Road engineering asphalt and asphalt mixture test procedure* (JTG 052-2000).

After demoulding, specimens kept in 40℃ ventilation oven by 72h curing. This curing method can simulate the condition after 10d actual pavement construction [8]. Splitting test: Specimens at 20℃ environment after 1.5 h do dry splitting test; Specimens at 20 ℃ water flooding after 24 h do wet splitting test. Using digital imaging microscope and SEM technology, we analysed the mechanism of increasing on the intensity of foamed asphalt mixture resulted by the addition of cement.

3 Results and discussions

3.1 Features of foamed asphalt

Fig. 1 is characteristic curve of foamed asphalt at three different temperatures. It is shown that at the same temperature, the foamed asphalt's half-life decreases with the increase of water consumption. The half-life decreases rapidly, when water consumption increases from 1.5% to 2.5%. But, from 2.5% to 3%, the half-life reduction is significantly smaller. The expansion ratio increases with the water consumption, and the increase is more uniform. Zhongtai AH-70 asphalt best foaming temperature is 160 ℃ and water consumption is 2%, and the expansion ratio is 11 times and the half-life is 10s.

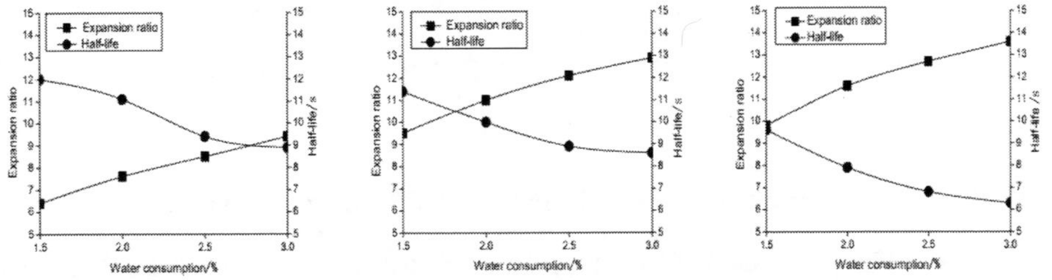

(a)150℃ (b)160 ℃ (c)170℃

Fig. 1. Characteristic curve of foamed asphalt at three different temperatures.

3.2 Design for proportion of cement-foamed asphalt cold recycled mixture

Table 3. RAP screening results and standard graduation

Graduation	Through rate of different sieve pore											
	26.5	19	16	13.2	9.5	4.75	2.36	1.18	0.6	0.3	0.15	0.075
RAP/%	100	98.9	95.7	84.9	65	48.7	34.4	24.0	13.3	3.11	0.92	0.24
Upper limit/%	100	100	-	-	85	65	55	-	-	30	-	20
Lower limit/%	100	90	-	-	60	35	30	-	-	10	-	6

Table 3 is RAP screening results and standard graduation. It is shown that the RAP's through rate of 0.3mm~0.075mm sieve pore are far below the standard. Fine aggregate is less than standard, especially the through rate of 0.075mm sieve pore is 0.24% that far from the 6% ~ 20% requirement in the standard. So proportion of the mixture is 75%RAP+25% fine aggregate. Fig. 2 is screening result of mixture that meets the requirement of standard graduation. Table 4 is the result of mixture's compaction test and mixing water consumption. Mixing water consumption based on the empirical formula of the Witgen cold regenerative technology manual: $W_{add}=1+(0.5W_{omc}-W_{air\text{-}dry})$. Wadd is mixing water consumption, W_{omc} is optimum moisture content and $W_{air\text{-}dry}$ is maximum dry density.

Table 4. The result of mixture's compaction test

Cement content/%	W_{omc} /%	$W_{air\text{-}dry}$ /%	W_{add} /%
0	5.92	2.212	3.96
1	6.13	2.229	4.05
1.5	6.22	2.236	4.10
2	6.31	2.243	4.13
3	6.49	2.256	4.21

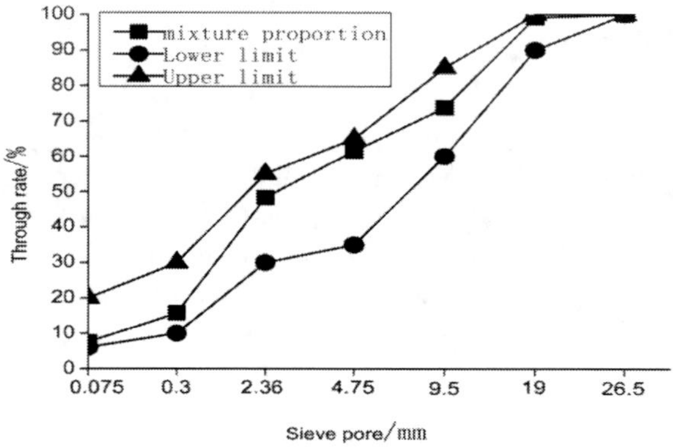

Fig. 2. Screening result of mixture.

3.3 The water stability of different cement content foamed asphalt mixtures

Fig. 3 is different cement content and foamed asphalt content mixture's splitting strength and Fig. 4 is wet splitting strength of the mixture. As shown in the Fig.s, the splitting strength of the mixture was increased significantly with the increase of the cement content, but the increase was gradually decreased. At the same cement content, there is the optimum foamed asphalt content. Under the condition of 2% cement content and 3% foamed asphalt content, splitting strength is 0.512 MPa and wet splitting strength is 0.416 MPa and dry and wet splitting strength ratio is 81.2%. It fully meets the standard requirement that the dry splitting strength is less than 0.4 MPa and the ratio is less than 75%. Therefore, for this mixture the best foamed asphalt addition is 3%, and proportion of the mixture is RAP: fine aggregate: cement=75:23:2

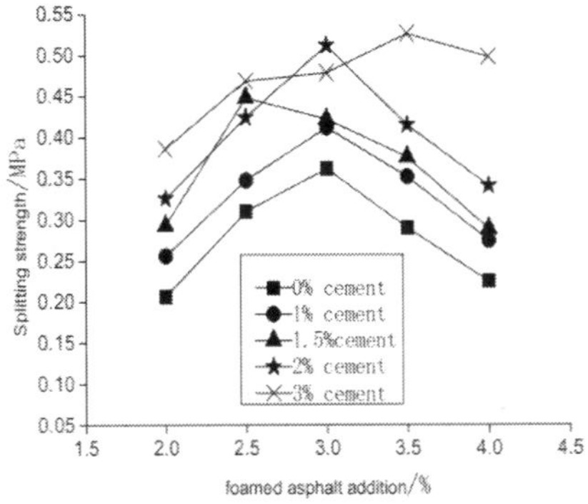

Fig. 3. Splitting strength of the mixtures.

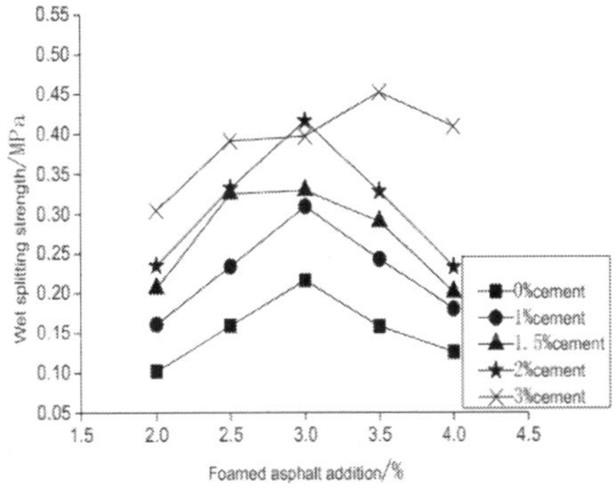

Fig. 4. Wet splitting strength of the mixtures.

After the hydration of the cement in cold cycling mixture, hydration products develop to the surrounding space of the cement particles, criss-cross, and gradually fill the mix of all the capillary space. Thereby it forms a uniform, dense; pore closed the whole structure, and enhances the overall strength of the mixture of water and anti-damaging [9]. It is clear that acicular and fibrous gels are interwoven overlapping to form a network structure in Fig. 5. Thus, the cement hydration products contained in the foamed asphalt mixture hydrolysed into space mesh structure and wrapped up the aggregate particle. It is the main reason that the cement can enhance the mixture's intensity as well as the water stability. The amount of cement is not the higher the better. The mixture of high cement adding is easy to produce severe shrinkage cracks and it leads to lower strength.

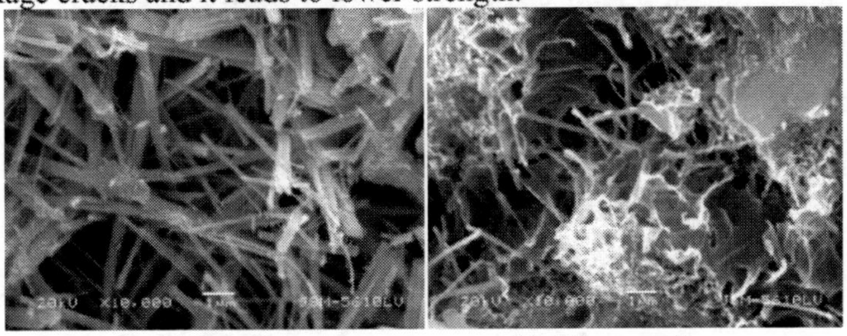

(a) Acicular gels (b) Fibrous gels

Fig. 5. SEM images for hydration products of the mixture

4 Conclusions

At the same temperature, the foamed asphalt ' s half-life decreases with the increase of water consumption, but the expansion ratio increases with the water consumption. Zhongtai AH-70 asphalt best foaming temperature is 160℃ and water consumption is 2%, and the expansion ratio is 11 times and the half-life is 10s.The splitting strength of the mixture was increased significantly with the increase of the cement content. At the same cement content, there is the optimum foamed asphalt content. For this mixture the best foamed asphalt addition is 3%, and proportion of the mixture is RAP: fine aggregate: cement=75:23:2.

This work was financially supported by National Key Research and Development Program (2016YFC1400802) and Provincial key Research and Development Program (2016GGH4501).

References

1. G.K. Wang, Z.L.Zhou. Foamed Asphalt Cold in-plant Recycling Applied in Pavement, Reconstruction Transport Standardization, **219** (2010),p.112-117
2. Guidelines for Cold In-Place Recycling[M]. Asphalt Recycling and Reclaiming Association, Annapolis, MD, 1991.
3. L.Y. Wei, Z.W. Liu, H.R Zhang, J. Wang. Experimental Analyse of Foamed Asphalt Stabilization Cold-recycled Mixture in Old Asphalt Pavement, Journal of Hebei University of Technology. **37**(2008), p.100-104
4. C.X Cao, G.P. He, C.R. Sun. The Cold Regeneration Technique by foamed asphalt, Highway, **11** (2003), p.99-101
5. Z.K. Xu. Effect of Foam Asphalt and Cement Content on Road Performance of Cold Recycled Asphalt Mixture, Highway Engineering, **42** (2017), p.245
6. T.Q. Ling, L. He, Y. Ma. A Cold Recycling Technique for Foamed Bitumen Mix, Journal of Civil Architectural & Environmental Engineering, **31** (2009), p.141-145
7. F.Z. Shi, X.J. Li, D.Q. Sun, W.M. Lv. Discussion about Design Methods for Recycling Cold Mix Asphalt [J].Highway, **11** (2004), p.103-107.
8. X.J. Li. Guide Experimental Study on Shearing Properties of Cold-Recycled Mixtures Using Foamed Asphalt, Journal of Building Materials,**13**(2010), p.27
9. X.B. Yin. Performance of Emulsified Asphalt Full Thickness Cold Recycling Subgrade Materials, Advanced Materials Research, **915-916** (2014),p. 704

Combined Effect of Initial Curing Temperature and Crack Width on Chloride Penetration in Reinforced Concrete Beams

Lotfi Elkedrouci[1*], *Bo* Diao[1, 2], *Sen* Pang[1], and *Yi* Li[1]

[1]School of Transportation Science and Engineering, Beihang University, Beijing 100083, China
[2]State Laboratory of Subtropical Building Science, South China University of Technology, Guangzhou 510640, China

Abstract. Reinforced concrete (RC) structures are gradually being degraded all over the world, largely due to corrosion of the embedded steel bars caused by an attack of chloride penetration. Initial curing would be regarded as one factor influencing chloride diffusion in concrete in combination with cover cracking that is also of great attention for reinforced structures. In this study, a non-steady state diffusion test of chloride ion involving RC beam specimens with a water-to-cement ratio of 0.5, initial curing temperatures of 5°C or 20°C and three types of crack widths ranging from 0 to 0.2mm was performed. Chloride content at 5°C or was determined. The results show that the higher chloride content was obtained in condition of crack width large than 0.1mm with low initial curing temperature and there are no obvious differences in chloride content when the crack width was not larger than 0.1mm.

1 Introduction

A durable reinforced concrete structure must withstand the couple actions of environment and loads over its service life (physical, chemical, mechanical...), as well as various actions such as wind, rain, cold, heat, ambient... while maintaining its aesthetic. Chloride penetration is one of the frequent aggressive agents affecting durability of reinforced concrete structures by corrosion of steels bars, which leads to the deterioration and even collapse of concrete structures [1]. One of the predominant ways that chloride penetrate into concrete structures is diffusion [2], which can be affected by several important factors. During service life, concrete structure may suffer complicated stress conditions of compression or tension leading to inevitable cracks due to the weakness of concrete in tension, which have a deep effect on chloride diffusion [3] and may results in different crack widths distribution within the change of concrete under different external loads, causing an easy way to chloride diffusion [4]. Incursion of chloride ion into concrete leads to break the passive film causing steel corrosion and resulting in cracking and spalling of concrete cover [5], in addition it also leads to decreasing the strength of reinforced

*Corresponding author: chinepekin0@gmail.com

concrete structures and the structural safety [6]. Obviously, existence of crack has a significant effect on chloride ingress which leads to increase the chloride diffusion [7]. Many researchers have studied chloride diffusion through concrete structures separately without considering the couple effects of an external load and aggressive agent in service environment [8]. The initial curing condition to which concrete is subjected has a significant effect in determining the chloride diffusion rate in concrete [9]. Poor initial curing conditions cause drying of cementitious materials especially at the surface, leading to restrict the hydration at the surface layer as well as the higher porosity and permeability [10]. The lower hydration of cementitious materials may happen in lower curing temperature and accelerated in higher curing temperature [11].

In this study, the combined effects of initial curing temperature and loading crack width on chloride permeation have been experimental investigated. The chloride content was tested and analyzed according to variations of initial curing temperature and loading crack width. The non-steady state accelerated migration test was employed to explore chloride concentration.

2 Experimental study

2.1 Materials and specimens design.

Two groups, total of 54 reinforced concrete beam specimens were cast for experiment using 42.5 ordinary Portland cement with water/cement ratio 0.5, where 27 beam specimens for each group. Two reinforcing steel bars of 340 mm in length and 10 mm in diameter were placed in tensile area of reinforced concrete beam. Two stainless steel electrodes were employed for electrical migration test measurements. They were attached at the extremities of the steel bars, which were impregnated at a depth of 30mm as shown in Fig. 1(a).

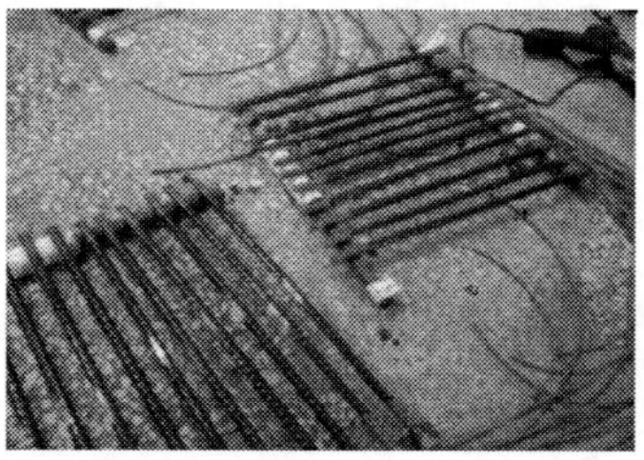

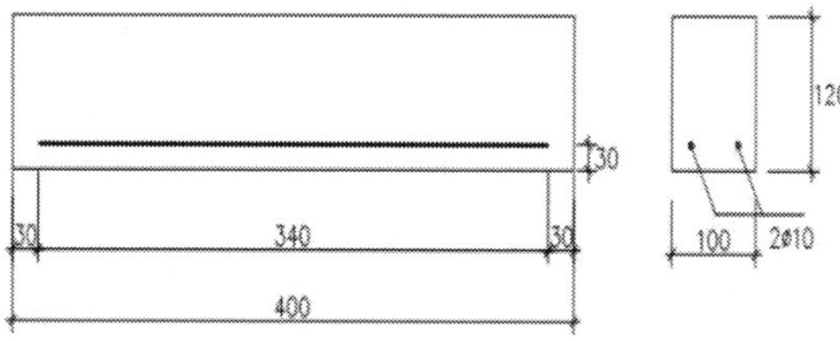

(a) (b)

Fig. 1. Reinforcement bars and geometry size of beam specimen, (a): Steel bars, (b): Geometric sizes of beam specimen.

Reinforced concrete beam specimens were made in factory of prefabrication. The concrete beam specimens were cast inside steel molds, then compacted on a vibrating table. The concrete beam specimens were demolded after 24h.

2.2 Curing conditions

After demolding, two groups of beam specimens were cured in two initial curing temperatures of 5°C or 20°C for 28 days, respectively. First group of beam specimens was cured in normal temperature of laboratory, which was simply controlled about 20°C, this normal temperature group was named NT for short. Second group of concrete beam specimens was initially cured in a low temperature chamber of 5°C, this group was named LT for short.

2.3 Initial cracks by three-point loading

The chloride contents of three points bending test was experimental investigated by several researchers such as [12]. However, there are no experiments deal with the initial curing temperatures. The concrete beam specimens of 400mm length, 100mm wide and 120mm height here were designed. And flexural cracks were produced by applying three-points loading. As soon as bending crack appeared, unload the beam specimens and crack were closed. Then the crack widths were inspected and photographed by crack width measuring instrument. The different widths of flexural cracks were performed by static three-point loading as illustrated in Fig. 2.

(a)

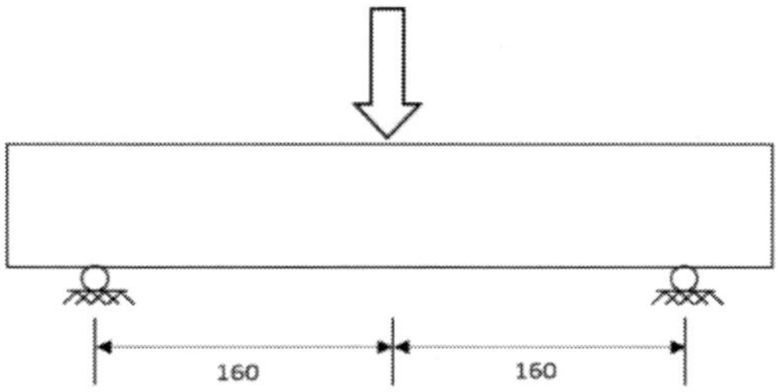

(b)

Fig. 2. Three-points loading of reinforced concrete beam specimen, (a): Photo of three-points loading, (b): Schematic diagram of three-points loading.

According to the crack widths, the beam specimens were classified into three types in this study. First type, crack widths were wider (in the range of 0.1mm~0.2mm) labelled 'W'. Second type, crack widths were narrow (in the range of 0.05mm~0.1mm) labeled 'N'. Third one, as reference, the beam specimen without crack labelled '0'.

2.4 Accelerated chloride migration test

In order to explore the effects of initial curing temperature and width of concrete crack on chloride diffusion, a non-steady state accelerated migration test method was employed and developed in our laboratory basing on forcing chloride ion to penetrate through cover using electrical field. The test method is similar to NT BUILD 492 developed by tang [13] which was investigated by many authors [14]. The principle of the test consists of sandwiching concrete beam specimens between two electrolyte compartments. The test was carried out based on bonding a PVC pipe of 60 mm diameter and 40 mm high on the cracked surface of the beam specimens) and sealed all the other cracked areas using epoxy resin to guarantee a one-way diffusion. Concrete specimens were submerged in pure water except the area which is covered by the PVC pipe for 3 to 5 days to ensure that concrete is saturated of water. an external electrical potential of 20–30V was applied for 24 hours, applied using a steel tablet immersed in chloride solution as a anode and the steel bars as a cathode as shown in Fig. 3. All beam specimens with different crack widths and under different initial curing temperatures were performed the same accelerated migration test.

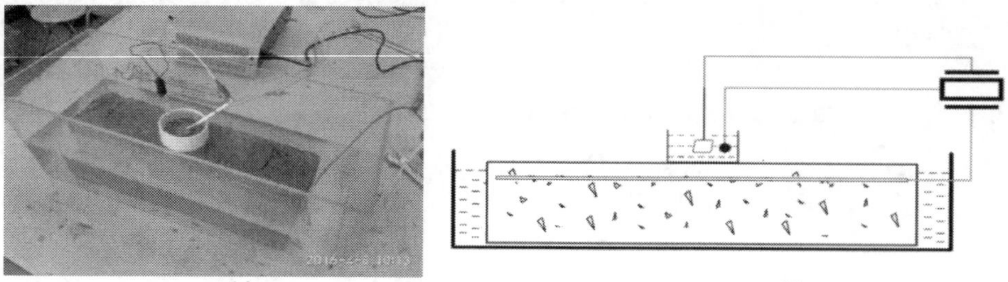

(a) (b)

Fig. 3. Schematic diagram of RCM test of RC beam, (a): RCM test of RC beam specimen, (b): Sketch of RCM test of RC beam specimen.

2.5 Chloride content test

After accelerated migration test completed, as shown in Fig. 4, concrete core samples of tensile cracked cover were drilled on one-way diffusion area to evaluate the chloride penetration. The height of core samples was the cover sickness of 30mm, each core sample was then cut into five circle slices along the depth of cover, the thickness of each circle slice was about 6mm, then circle slices were grinded into powder to measure the chloride ion concentration. Based on the results, chloride ion content in different depth can be measured.

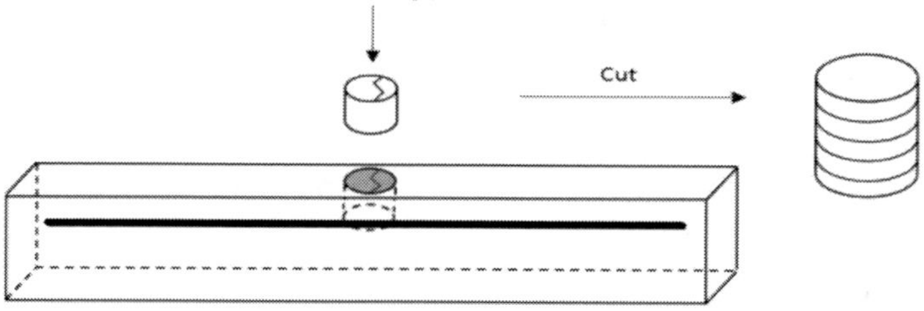

Fig. 4. Schematic diagram of core sample drilling and circle slices sub-cutting.

3 Results and discussion

3.1 Chloride content with crack width

A remark is to be made concerning the chloride content profiles of concrete beam without crack labelled '0', with narrow cracks (width 0~0.1mm) labelled "N" and with wider cracks (width 0.1~0.2mm) labelled "W". The variation of chloride content along the depth of concrete cover under two initial curing temperatures with three types of crack widths were shown in Fig. 5.

The variation tendency of chloride content for the three crack widths in both groups of initial curing temperatures was nearly same. High chloride content values were observed near the exposure surface of the beam specimens, then decrease dramatically at 10 mm of depth. Beyond 10-15 mm of depth reaching the surface of steel bars, the decreasing rate of chloride content is insignificant. For normal initial curing temperature, the chloride content of "W" beam specimens were the largest along the depth for three types of crack width beams, which means that the effects of crack width on chloride content was significant. For low initial curing temperature, the effect of crack width on chloride content was also obvious, where the chloride content was the largest when the crack width was larger than 0.1mm. As shown in Fig. 5, an obvious increase of chloride content was exhibited when the crack width of beam specimens was larger than 0.1mm (in the range of 0.1~0.2mm), no matter the initial curing temperature was lower or not.

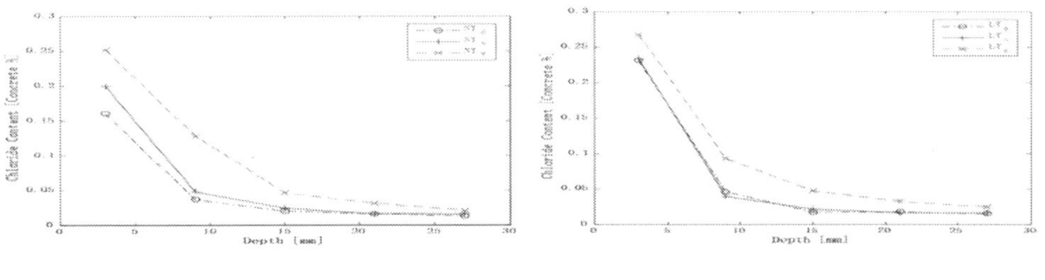

(a) (b)

Fig. 5. Chloride content with depth and crack widths for different initial temperatures, (a): Chloride content with depth and crack for NT, (b): Chloride content with depth and crack for LT

3.2 Chloride content with initial curing temperature

Fig. 6 shows the comparison of chloride content with depth at the same crack width for both initial curing temperatures. The chloride content for the three considered crack widths have the same tendency, as the depth increases, the value of chloride content presents no significant difference for both initial curing temperatures. Generally, crack provide the flow paths which leads to the increment of concrete permeability, as a result, aggressive agent or water can easily penetrate into the concrete and to corrode steel bar, leading to performance deterioration [15]. From exposure surface, the chloride content of beams whose crack width is 'W' is much larger than that of 'N' and '0' either in NT or LT. Concrete beam specimens with crack width '0' in both initial curing temperatures have almost the same chloride content values as concrete beam specimens with crack width 'N', this indicate that crack widths below 0.1mm doesn't have a significant effect as much as crack widths in the range of 0.1~0.2mm.

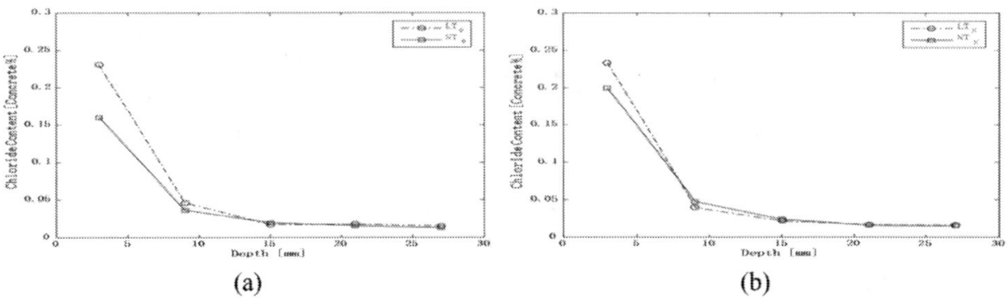

(a) (b)

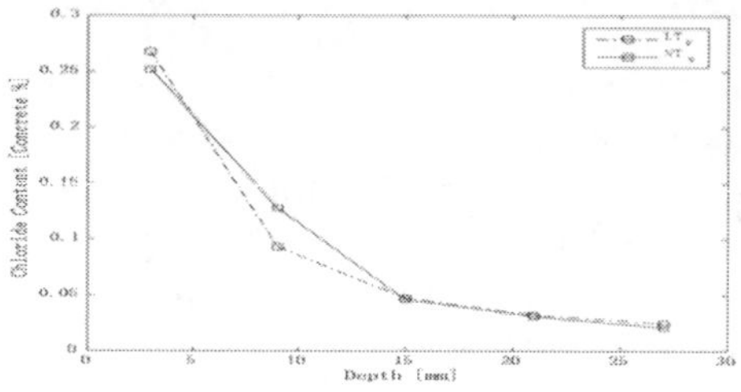

(c)

Fig. 6. Chloride contents along depth for three types of cracks widths and two initial curing temperatures, (a): Chloride content without crack, (b): Chloride content for narrow crack, (c): Chloride content for wide crack.

3.3 Chloride content at the surface of steel bar

The chloride content at the surface of steel bar for three types of crack widths and two initial curing temperatures were shown in Fig. 7. As shown in Fig. 7, at steel bar surface, concrete beam specimens with crack width 'W' exhibit the highest chloride content in all types of crack widths, which may lead to early corrosion of steel bar [15]. And the chloride contents of the concrete beam specimens curing initially in LT (Lower Temperature 5°C) are higher than that of concrete beam specimens curing initially in NT (Normal Temperature), this can be explained by the fact that initial curing in LT, leads to insufficient hydration process in the active components of the cement [11]. As shown in Fig. 7a, the wider of crack width, the higher chloride content for the same initial curing temperature.

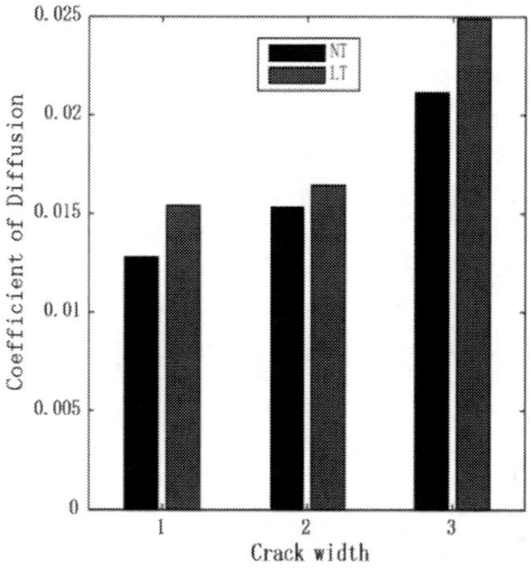

(a)

(b)

Fig. 7. Chloride content at the steel bar surface with crack width for LT and NT condition. (a): Chloride content with different crack widths, (b): steel bar surface after drilling core sample.

4 Conclusions

Based on the experimental results, the following conclusions may be concluded:

Chloride content of concrete with different cracks is highly dependent on the crack width. High values of chloride content were measured in concrete beam specimens in condition of crack width larger than 0.1mm and much less in concrete beam specimens with narrow crack (less than 0.1mm) and without crack ('0'mm) for both initial curing temperatures. This shows that wider crack results in high chloride penetration.

The chloride contents of RC beam specimens were almost the same values when the crack width was smaller than 0.1mm for both initial curing temperatures. This shows that narrow crack width (less than 0.1 mm) have almost the same effect such as concrete beams without crack (0mm) on the chloride penetration.

The chloride contents of cracked RC beam specimens cured initially at 5°C were higher than those at 20°C. This indicates that initial curing temperature have significant impact on chloride penetration of cracked RC beam specimens.

This work is part of the projects financially supported by the National Natural Science Fund (NSF) of China Grant No. 51678021 and the Fund of State Key Laboratory of Subtropical Building Science (FSKLSBS), South China University of Technology Grant No. 2016ZA03. Authors gratefully acknowledge the funding from the NSF and FSKLSBS.

References

1. B.H. Oh, S.Y. Jang, cem.con.res, **34** 463–480 (2004)
2. L-O. Nilsson, L. Tang (Springer, NL, 289-311 1996)
3. S. Y. Jang, B. S. Kim, B. H. Oh, Cem. Con. Res, **41**, 1 (2011)
4. M. Ismail, A. Toumi, R. François, R. Gagné, Cem. Con. Res, **38**, 8 (2008)
5. Z. Shao-feng, L. Chun-hua, L. Rong-gui, Pro.E, **24** (2011)
6. C. Lu, W. Jin, R. Liu, Corr. SCI, **53**, 4 (2011)
7. H. Wang, W. Jin, C. Lu, Y. Bai, J. Mat. Civ. E, **23**, 7 (2011)
8. R.K. Dhir, M.R. Jones, A.E. Elghaly, Cem. Con. Res, **23**, 5 (1993)
9. C. Jaegermann, Mat. J, **87**, 4 (1990)
10. G. C, Mays (Concrete Press, 2002)
11. J.-K. Kim, S.H. Han, Y.C. Song, Cem. Con. Res, **32** (2002)
12. T. U. Mohammed, N. Otsuki, M. Hisada, T. Shibata, J. Mat. Civ. E, **13**, 3 (2001)
13. T. Luping, L.-O. Nilsson, ACI. Mat. J, **89**, 49–53 (1993)
14. O. Truc, J.P. Ollivier, M. Carcassès, Cem. Con. Res, **30**, 217–226 (2000)
15. K. Wang, D.C. Jansen, S.P. Shah, A. F. KArr, Cem. Con. Res, **27**, 3 (1997)

The Effect of Mixing and Curing Sea Water on Concrete Strength at Different Ages

Qingyong Guo[1] *, *Lei* Chen[1], *Huijian* Zhao[1], *Jorge* Admilson[1], and *Wensong* Zhang[2]

[1]College of Aerospace and Civil Engineering, Harbin Engineering University,Harbin 150001, China
[2]General Research Institute for Nonferrous Metals, Beijing 100088, China

Abstract. In the research, the effects of sea water for mixing, curing on the gain in strength of different grades of concrete was investigated. A total of 192 concrete cubes were tested for their compressive strength. The study shows that sea water affects the rate of gain in strength of concrete when used for mixing or curing. The strength of concrete made by using sea water was observed to be decreased by about 15% as compared to the similar concrete specimens made and cured with fresh water at 90 days. The concrete with higher strength showed poorer resistance against strength deterioration as compared to the lower strength concrete which used sea water for curing. And the concrete made with sea water decreased the stability of concrete properties.

1 Introduction

The United Nation estimates that till 2050 the world will get an additional population of about 2.5bilion [1], with the largest growth in developing countries that have experienced in water crisis. In addition more than half population will not be able to get enough drinking water. Increasing infrastructure development is in line with the needs of clean water every year, where sea water has not been used for such things for that reason. Study of fresh water savings is very seriously needed. The use of sea water, the percentage is 97% from the total water on the earth, is absolutely necessary.

In coastal areas, there has always been a deficiency of fresh water as the available water is affected by sea salts. So it is difficult to deliver fresh water for construction works in such location. Also it is economical to use sea water that is available near the construction site instead of potable water to be transported from other areas sources. But sea water contains large amount of sea salts, which may have unfavorable effect on the properties of concrete[2]. Till now, there are many researchers investigated on the strength behavior of concrete made with plain as well as sea water and also to be exposed in marine condition[3-7].

Although the existing literature and codes of practice reveal the effect of mixing and curing of sea water on durability of concrete, it still remains an area requiring further study and research particularly of the use of structural concrete in marine environment.

* Corresponding author: qingyong622@126.com

The main purpose of this research is to investigate the potential of replacing the fresh water by salt sea water in concrete mixtures and identify its mechanical behavior when mixed and cured with salt sea water. So in this study, the strength behavior of concrete made and cured with sea water on the gain in strength of different grades of concrete is investigated.

2 Experimental program

The experimental program was planned to investigate the effect of sea water for mixing and curing on the gain in strength of concrete at different ages. This study includes determination of compressive strength (up to 90 days) of 192 specimens made and cured by using plain water as well as sea water.

2.1 Materials used

The details of various materials used in the experimental investigation are as following:

(1) Cement: The cement used was Ordinary Portland Cement (OPC) 42.5 grade. It was stored under dry condition, free of lumps and in conformity with GB/T50081-2011[8].

(2) Aggregates: The coarse aggregate used was crushed stone with a size of 15-20 mm. The fine aggregate used was medium sand having a particle size greater than 0.25 mm and a mass exceeding 50% of the total mass. Fineness modulus was in the range of 3.0-2.3.

(3) Mixing Water: The water used was fresh and sea water, the fresh water was gotten from the tap in the laboratory, as well was the sea water gotten but mixed with sea salt.

2.2 Test layout and procedure

(1) Concrete quality: Four different groups of concrete, namely concrete A, B, C and D were used. Relevant information of the concrete mixes is given in table 1, where each group were mixed and cured in fresh water (FF), mixed and cured in sea water (SS), mixed with fresh water and cured in sea water (FS), and mixed with sea water and cured with fresh water (SF). Sea water means water mixed with salt, and fresh water means tap water or plain water.

(2) Exposure period: Test specimens were tested periodically after specified curing periods of 90 days in fresh water as well as sea water.

(3) Size of specimen: Concrete cube of 100×100×100(mm) was used for measuring the compressive strength determined according to GB/T50081-2011[8].

(4) Curing environment: A total of 192 concrete cubes were cast in the laboratory. And then the specimens were kept at 22°C temperature for 24 hours. After demolding, all specimens were made and cured as per program schedule indicated in table 1.

Table 1. Details of Concrete Mixes

Concrete Category	W/C Ratio	OPC	Fine aggregate	coarse aggregate	Mixing and curing mode	Exposure Periods (Days)
A	0.50	1	1.5	3	FF	7
B	0.45	1	1.5	3	FS	14
C	0.50	1	2	4	SF	28
D	0.45	1	2	4	SS	90

Note:FF-Concrete cast and cured with fresh water; SF - Concrete cast with sea water and cured with fresh water ; FS - Concrete cast with fresh water and cured with sea water; SS - Concrete cast and cured with sea water.

Compressive strength test was performed on the concrete cubes at the curing age of 7, 14, 28 and 90 days using the compressive testing machine. After the specific period of exposure, the cubes specimen were compressed at uniform rate 4.0 kN/sec (without shock) until failure occurred[9].

3 Test results and discussion

The result was the average strength of three standard-cured specimens made from the same concrete sample and tested at the same age. The compressive strengths at the age of 7, 14, 28 and 90 days for all groups were represented in table 2 and figures below.

3.1 Variation of compressive strength with ages

According to the variation of compressive strength in Fig. 1 to Fig. 4, it was observed that the concrete made with sea water (SS and SF) showed rapid increase in early strength, compared with the concrete made with fresh water(FF and FS) at 7 days of curing. The early increase in strength of concrete SS and SF are basically due chlorides contains in sea water that tends to accelerate the setting of cement and to improve the strength. While the rate of gain in strength of sea water mixed concrete (SS and SF) and sea water mixed and fresh water cured (SF) are observed to be less than that of fresh water mixed concrete (FF and FS) at 14d, 28d and 90d, due to leaching out of soft hydration product or the sulphates contains in sea water that retard the setting of cement.

Table 2. Average compressive strength

Concrete mix	Concrete designation	Compressive strength (MPa)			
		7 days	14 days	28 days	90 days
A	FF	39.47	45.18	47.17	59.31
	FS	42.05	46.20	49.79	54.98
	SF	39.10	41.47	41.40	45.38
	SS	38.03	43.63	46.14	53.31
B	FF	41.17	46.12	54.33	65.25
	FS	41.30	46.46	49.98	62.51
	SF	43.66	46.46	45.37	57.14
	SS	42.18	45.51	48.32	54.64
C	FF	36.60	48.15	49.02	62.00
	FS	41.32	47.42	49.70	56.05
	SF	40.99	43.24	41.55	53.64
	SS	40.98	44.11	46.46	56.63
D	FF	46.24	47.04	59.09	66.95
	FS	42.32	49.13	58.56	63.20
	SF	37.85	46.11	44.98	46.36
	SS	44.79	52.37	50.69	57.37

It is shown that the sea water has significant effects on the gain of concrete strength with curing age. Sea water for mixing and curing concrete results in around 15% of strength loss compared to the concrete made and cured with fresh water. Fresh water mixed and sea water cured concrete also shows losses in strength comparing to the control batch (FF) around 7%.

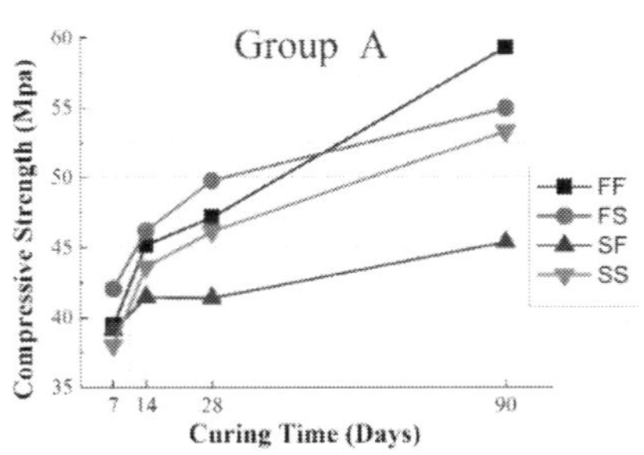

Fig. 1. Compressive strength for group A

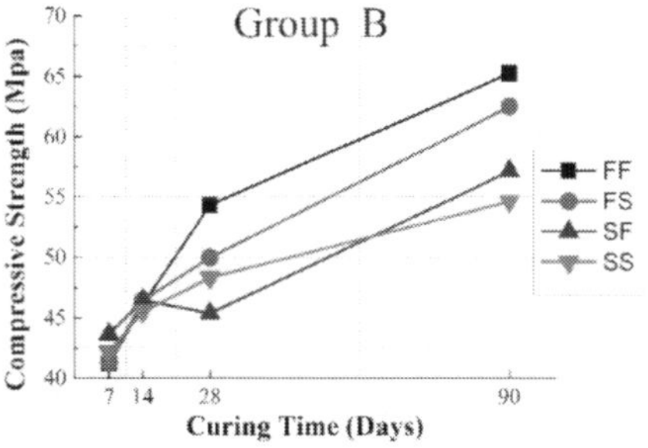

Fig. 2. Compressive strength for group B

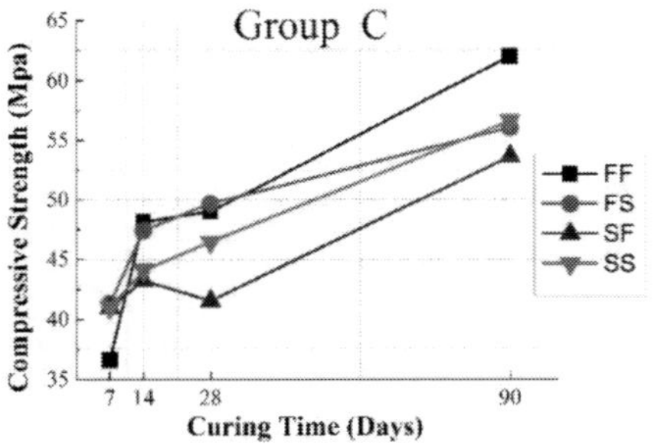

Fig. 3. Compressive strength for group C

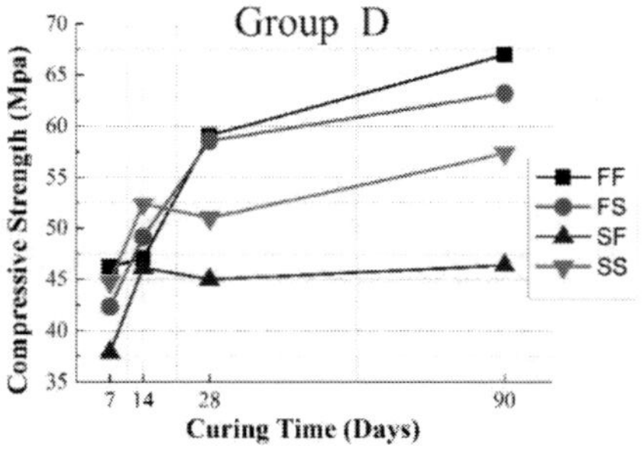

Fig. 4. Compressive strength for group D

Comparing the four types of concrete designation FF and SS above, it is seen that, strength losses of concrete is observed to vary with the grade of concrete. At curing age 28 days, the overall losses of strength of concrete SS are observed to 1.03Mpa for group A, 2.56MPa for group C, 6.01MPa for group B and 8.40MPa for group D compared to the concrete FF. Then at curing age 90 days, the losses are observed 5.37MPa for group C, 6.0MPa for the group A, 9.58MPa for group D and 10.61MPa for group B. Among all the concretes studied, Group D and B are found to be mostly affected by the sea water. The possible cause for strength deterioration may be due to dissolution of the compounds rich in lime and the formation of expansive compounds as a result of chloride and sulfate attack of sea water.

So it can be concluded that relatively higher strength concrete showed poorer resistance against strength deterioration as compared to lower strength concrete. In opposition, sea water mixed and cured concrete shows around 15% higher strength loss as compared to fresh water mixed and cured concrete.

3.2 Variation of relative intensity with concrete grade

In order to be able to more closely reflect the law of compressive strength versus age, the relative compressive strength at different ages of four concrete batches were calculated using the ratio of strength at each age with the intensity of 28 days. As seen in the relative intensity graphic as shown in Fig. 5 to Fig. 8 below, the four concrete designation graphics assembled in one graphic for four concretes designation are dispersed around the logarithm line of the function equation y, taking the concrete designation FF as standard result, which according to the variable $x = f_n / f_{28}$ .The correlation coefficients (R^2) were calculated whether the closer it gets to 1 the better it is.

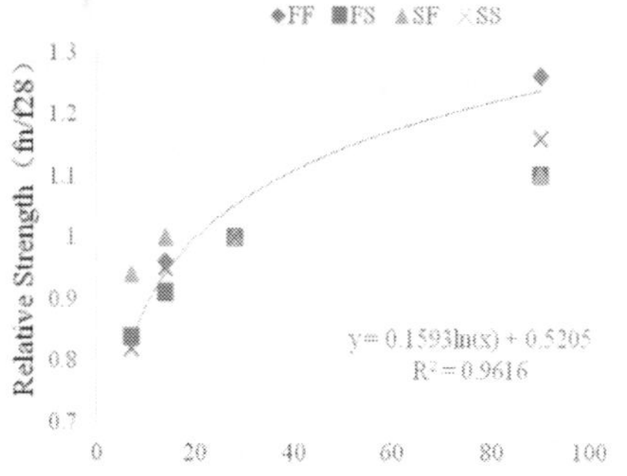

Fig. 5. Relative intensity for group A

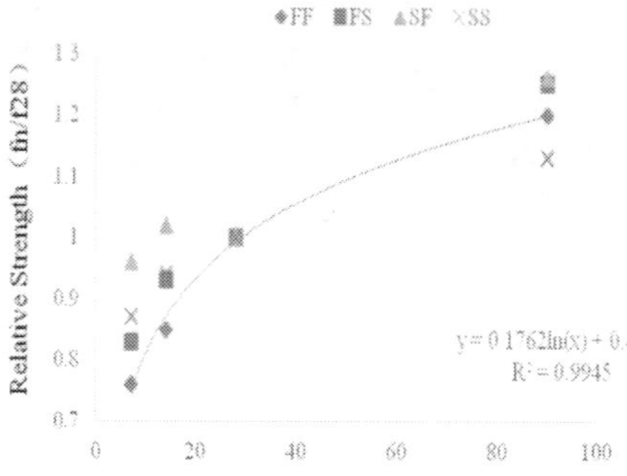

Fig. 6. Relative intensity for group B

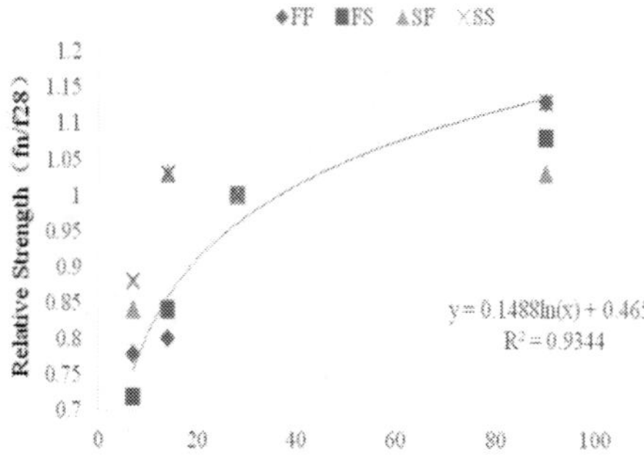

Fig. 7. Relative intensity for group C

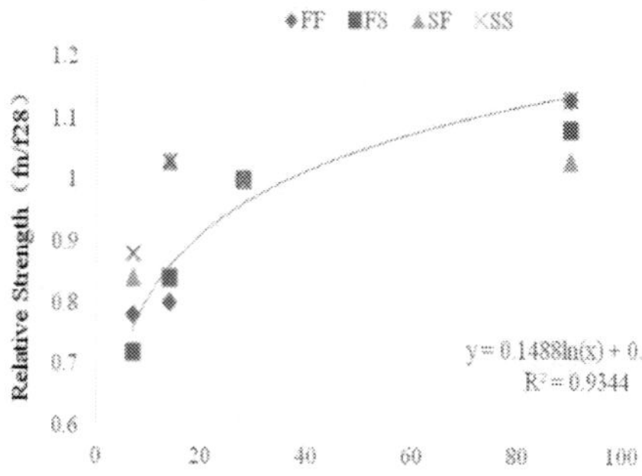

Fig. 8. Relative intensity for group D

It is found that the concrete made with sea water (SS and SF) showed bigger dispersion than that of the concrete made with fresh water (FF and FS). It means that the sea water decreases the stability of concrete properties. Therefore for application some precautions and analysis of the effect of sea water in mixing and curing concrete has to be taken.

4 Summary

The results of the research investigated on four different grades of concrete made with fresh and sea water, cured in fresh water as well sea water over a period of 90 days for compressive strength have been analyzed. The following conclusions are drawn:

(1) Sea water affects the gain in strength of concrete when used for mixing and curing. It shows some increases at the early strength but ultimately decrease the strength.

(2) Concrete specimens made with fresh water and cured with sea water shows a loss in strength of around 7% whereas concrete specimens made and cured with sea water showed loss in strength of around 15% as compared to the similar concrete specimens made and cured with fresh water at age 90 days.

(3) Among the four grades of concrete studied, the higher grade concrete B showed around 9% higher strength loss as compared to the other grades of concrete A, C and D when used sea water for mixing and curing concrete. And the higher strength concrete showed poorer resistance against strength deterioration as compared to the lower strength when used sea water for curing.

This research was financially supported by Natural Science Foundation of Heilongjiang Province of China(NO: E201415) and the Fundamental Research Funds for the Central Universities of China (No.GK2020260143).

References

1. G. Tchobanoglous, F. L. Burton, and H. D. Stensel. *Wastewater Engineering: Treatment and Reuse(4th ed)* (Metcalf & Eddy Inc., New York,2003)
2. Ghorab H. Y. Cement and Concrete Research. **19**,868 (1989)
3. Olutoge, F. Adeyemi, Amusan, G. Modupeola.International Journal of Engineering Science Invention. **3**, 23 (2014)
4. Akinkurolere O. O., Jiang C, Shobola O. M. Journal of Engineering and Applied Sciences. **2** ,412 (2007)
5. H.K. Liu. N.H. Tai. W.H. Lee. Composites Science and Technology. **62**,2131 (2002)
6. Uddin, T., Hidenori, H. and Yamaji, T. Cement and Concrete Research. **34**, 593 (2004)
7. Zaher, K. and Shihada, S. Journal of the Islamic University of Gaza.**11**,156 (2003)
8. Chinese Standard. *Common Portland Cement standard*(Standards Press of China, Beijing, China, 2015)
9. Chinese Standard. *Standard for technical requirements and test method of flexural strength* (Standards Press of China, Beijing, China, 2011)

Performance on Water Stability of Cement-Foamed Asphalt Cold Recycled Mixture

Junxiao Li [1*], *Wei* Fu [2], and *Hechao* Zang [1]

[1]Shandong Provincial Key Laboratory of Ocean Environmental Monitoring Technology, Institute of Oceanographic Instrumentation, Shandong Academy of Sciences, Qingdao 266001, China
[2]Navy Qingdao First Sanatorium, Qingdao 266001, China

Abstract. Through designing the mixture proportion of foamed asphalt cold in-place recycled mixture combined with the water stability experiment, it shows that the addition of cement can obviously improve foamed asphalt mixture's water stability and the best cement admixture is between 1% ~ 2%; Using digital imaging microscope and SEM technology, the mechanism of increasing on the intensity of foamed asphalt mixture resulted by adding cement was analyzed. It revealed that the cement hydration products contained in the foamed asphalt mixture hydrolyzed into space mesh structure and wrapped up the aggregate particle ,this is the main reason that the cement can enhance the mixture's intensity as well as the water stability. This research provides reference for cement admixture's formulation in the designing of foamed asphalt cold in-place recycled mixture.

1 Introduction

By the end of 2016, the total number of highways in china was 130,000 kilometers. So there will be a large amount of recycled asphalt mixture in the road pavement renovation, renovation and conservation every year. If calculating by 12% of highway pavement need repair, there will be 2.2 million tons recycled asphalt mixture every year[1]. Asphalt cold recycling can be a good solution, which can make the old pavement materials reused and save 40%-50% of the total investment[2]. Foamed asphalt is the bituminous material that contains a large amount of evenly dispersed bubbles after expansion. In a special foaming device, the expansion occurs when the heat asphalt comes into contact with the water. Now, foamed asphalt is used as a regenerative agent for recycled RAP. When foamed asphalt comes into contact with the aggregate, it suddenly becomes millions of tiny particles and spreads on the surface of the fine aggregate especially particle size less than 0.075 mm [3,4].This kind of the fine aggregate with foamed asphalt can be filled in the gap between the coarse aggregate and after mixing and compacting, the asphalt rubber paste is formed that make the mixture a certain cohesion and strength. In recent years, the cold regeneration technique of recycled waste road materials by foamed asphalt has been famous for its remarkable advantages in energy saving, environmental protection and economy[5,6,7].The

* Corresponding author: caike0501@163.com

effect of foam asphalt cold regeneration is influenced by the amount of active filler such as cement. In this report, the influence factors of the stability performance of different cement foamed asphalt cold in-place recycled mixtures were studied by splitting strength test. Furthermore, using digital imaging microscope and SEM technology, we analyzed the mechanism of increasing on the intensity of foamed asphalt mixture resulted by the addition of cement.

2 Experiment

2.1 Raw materials

The recycled asphalt pavement (RAP) is from highway milling material in Wuhan, China. Through foaming experiment of Zhongtai AH-70 asphalt, the best foaming temperature is 160℃ and water consumption is 2%, and the expansion ratio is 11 times and the half-life is 10s. Stone dust is 0 ~ 3mm limestone. The mine powder is limestone powder. Cement is P.O. 425 ordinary Portland cement from Huaxin cement factory, and table 1. is its chemical composition. Water is just general water.

Table 1. Chemical composition of cement.

Chemical composition	SiO_2	Al_2O_3	Fe_2O_3	CaO	MgO	Na_2O	SO_3	Ignition loss
Content /%	21.73	4.25	2.67	63.59	2.21	0.55	2.39	1.75

2.2 Specimen preparation and test methods

By comparing the particle size distribution of RAP with engineering design level gradation of foamed asphalt cold in-place recycled mixture, the mixture gradation is determined. Foamed asphalt cold in-place recycled mixtures have five different cement compositions of 0%, 1%, 1.5%, 2%, 3%. The optimized amount of water used and the maximum dry density were determined by the compaction test, and the mixing water consumption is calculated. The impaction test was referring to the method in <Highway Engineering stable inorganic binder materials testing procedures> (JTG E 51-2009). The preparation of mixture standard Marshall specimens was refer to the methods in < Road engineering asphalt and asphalt mixture test procedure > (JTG 052-2000).

After demoulding, specimens kept in 40℃ ventilation oven by 72h curing. This curing method can simulate the condition after 10d actual pavement construction [8]. Splitting test: Specimens at 20℃ environment after 1.5 h do dry splitting test; Specimens at 20 ℃ water flooding after 24 h do wet splitting test. Using digital imaging microscope and SEM technology, we analyzed the mechanism of increasing on the intensity of foamed asphalt mixture resulted by the addition of cement.

3 Results and discussion

3.1 Effect of Portland cement on the foamed asphalt mixture

Under the condition of 3% foamed asphalt, Table 2. is the splitting strength of five different cement composition specimens on different curing conditions. A is specimen curing condition after 24h 20℃. B, C and D are specimen curing conditions after 24h, 48h and 72h 40 ℃ .Under the condition of 3% foam asphalt, and based on the splitting strength of specimen that is kept in 40℃ ventilation oven by 72h, Fig. 2. is the percentage increase splitting strength of five different cement composition specimens on different curing conditions.

Table 2. Splitting strength of different cement composition specimens on different curing conditions.

Cement content /%	A splitting strength /MPa	B splitting strength /MPa	C splitting strength /MPa	D splitting strength /MPa	D wet splitting strength /MPa	D dry and wet splitting strength ratio /%
0	0.080	0.293	0.344	0.361	0.216	59.8
1	0.157	0.350	0.375	0.412	0.309	75.1
1.5	0.130	0.328	0.370	0.420	0.329	78.4
2	0.143	0.333	0.486	0.512	0.416	81.2
3	0.120	0.268	0.364	0.479	0.397	82.9

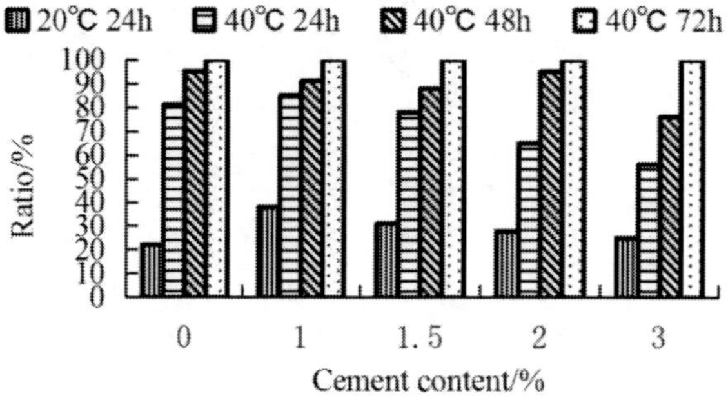

Fig.1. Increase splitting strength ratio of five different cement composition specimens on different curing conditions.

Based on the splitting strength of specimen that is kept in 40℃ ventilation oven by 72h, Fig. 1. shows that after 24 h at room temperature curing, the increase splitting strength ratio of the mixture specimen without adding cement is 22%, and the increase splitting strength ratio of the mixture specimens that adding cement is about 30%. It shows that splitting strength of different cement composition mixtures is lower. After 42h at 40℃ temperature curing, the increase splitting strength ratio of the mixture specimens that adding 1%~2% cement is about 80%, but the increase splitting strength ratio of the mixture specimens that adding 3% cement is only 55%. It shows that early curing stage for foamed asphalt cold in-place recycled mixtures of 1%~2% cement adding amount has rapid growth splitting strength and it is helpful to shorten the construction period of the road. Table 2. shows that dry and ret splitting strength ratio of foamed asphalt cold in-place recycled mixtures of 1% cement adding amount is 75.1% that is more than59.8% of none cement adding mixture. So, adding cement can obviously improve the water stability of foam asphalt mixture. When cement adding amount is 3%, dry and ret splitting strength ratio is 82.9% that is not much higher than 1% cement adding mixture. Based on engineering economy and dry and ret splitting strength ratio not less than 75% which is the rule for the regeneration of road asphalt pavement, the best cement admixture of this foamed asphalt cold recycled mixture is between 1% ~ 2%.

3.2 Mechanism

Fig. 2(a). is the digital imaging microscope images for 1% cement adding amount Marshall specimen surface. Fig. 2(b). is the digital imaging microscope images for 2% cement adding amount Marshall specimen surface. The figure shows that 1% cement adding amount Marshall specimens surface accumulation and crowding are compact and porosity is lower than 0% cement adding amount one. The greater porosity, the permeability is increasing. The moisture content is easy to infiltrate the interface of asphalt and the aggregate, and exchange caking property causing water damage. As a result, the stability of the mixture is reduced.

(a)1% cement adding amount (b) 2% cement adding amount

Fig.2. Digital imaging microscope images of the mixture

After the hydration of the cement in cold cycling mixture, some of the hydration products C_4AH_{13} react with gypsum: $C_4AH_{13}+3C\bar{S}H_2+14H=AFt+CH$. The needle bar AFt will be filled space within the mix and form a network structure that increased mixture strength. Others hydration products C-S-H coagulation forms a fibrous structure that further increased mixture strength [9]. Hydration products develop to the surrounding space of the cement particles, criss-cross, and gradually fill the mix of all the capillary space. Thereby it forms a uniform, dense, pore closed the whole structure, and enhances the overall strength of the mixture of water and anti-damaging [10]. It is clear that acicular and fibrous gels are interwoven overlapping to form a network structure in Fig. 3. Thus, the cement hydration products contained in the foamed asphalt mixture hydrolyzed into space mesh structure and wrapped up the aggregate particle. It is the main reason that the cement can enhance the mixture's intensity as well as the water stability. The amount of cement is not the higher the better. The mixture of high cement adding is easy to produce severe shrinkage cracks and it leads to lower strength. This research provides reference for cement admixture's formulation in the designing of foamed asphalt cold in-place recycled mixture.

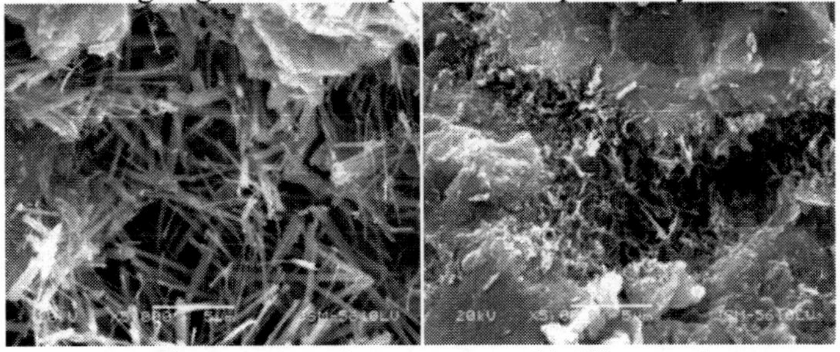

(a) Acicular gels (b) Fibrous gels

Fig.3. SEM images for hydration products of the mixture

4 Conclusions

Addition of cement can obviously improve foamed asphalt mixture's water stability and the best cement admixture of this foamed asphalt cold recycled mixture is between 1% ~ 2%. Early curing stage for mixtures of 1%~2% cement adding amount has rapid growth splitting strength and it is helpful to shorten the construction period of the road. The cement hydration products contained in the foamed asphalt mixture hydrolyzed into space mesh structure and wrapped up the aggregate particle. It is the main reason that the cement can enhance the mixture's intensity as well as the water stability. But the amount of cement should be controlled in the appropriate range. This research provides reference for cement admixture's formulation in the designing of foamed asphalt cold in-place recycled mixture.

This work was financially supported by National Key Research and Development Program (2016YFC1400802) and Provincial key Research and Development Program (2016GGH4501).

References

1. G.K. Wang, Z.L.Zhou. Foamed Asphalt Cold in-plant Recycling Applied in Pavement, Reconstruction Transport Standardization, **219** (2010), pp.112-117
2. Guidelines for Cold In-Place Recycling[M]. Asphalt Recycling and Reclaiming Association, Annapolis, MD, 1991.
3. L.Y. Wei, Z.W. Liu, H.R. Zhang, J. Wang. Experimental Analyse of Foamed Asphalt Stabilization Cold-recycled Mixture in Old Asphalt Pavement, Journal of Hebei University of Technology. **37**(2008), p.100-104
4. C.X. Cao, G.P. He, C.R. Sun. The Cold Regeneration Technique by foamed asphalt, Highway, **11** (2003), p.99-101
5. Z.K. Xu. Effect of Foam Asphalt and Cement Content on Road Performance of Cold Recycled Asphalt Mixture, Highway Engineering, **42** (2017), p.245
6. T.Q. Ling, L. He, Y. Ma. A Cold Recycling Technique for Foamed Bitumen Mix, Journal of Civil Architectural&Environmental Engineering, **31** (2009), pp.141-145
7. F. Z. Shi, X.J. Li, D.Q. Sun, W.M. Lv. Discussion about Design Methods for Recycling Cold Mix Asphalt [J].Highway, **11** (2004), p.103-107.
8. X.J. Li. Guide Experimental Study on Shearing Properties of Cold-Recycled Mixtures Using Foamed Asphalt, Journal of Building Materials,**13**(2010), p.27
9. Y.Z. Yuan. Gelling Material Science, Wuhan University of Technology Press, 2008.p.93-101
10. X. B. Yin. Performance of Emulsified Asphalt Full Thickness Cold Recycling Subgrade Materials, Advanced Materials Research, **915-916** (2014), p. 704

Preparation of Autoclaved Foamed Concrete Block from Fly Ash and Carbide Slag

Xing Tan[1], *Fuqiang* Han[1], and *Fengqing* Zhao[1, 2*]

[1]Hebei University of Science & Technology, Shijiazhuang 050018, PR China
[2]Hebei Engineering Research Center of Solid Wastes Utilization

Abstract. To achieve the comprehensive utilization of solid waste and reduce costs, fly ash, carbide slag, and low-clinker cement were used to produce lightweight foamed concrete block. Granulated blast-furnace slag (GBFS) was used as composition correction material in the block. The effects of curing temperature and dosage of low-clinker cement on the performance of foamed concrete block were investigated. The optimal material proportioning is obtained: fly ash 58.5%, carbide slag 20%, GBFS 10%, gypsum 1.5% and low-clinker cement 10%. The proper curing regime is "temperature rising 4h-180℃ constant temperature 4h-natural cooling". The results indicate that the compressive strength of the block reaches 3.55 MPa while the density is 616.9 kg/m^3. The performance of the product meets JC/T 1062-2007 (China professional standard of foamed concrete block).

1 Introduction

Wall materials innovation and energy saving in building are important parts in implementing sustainable development strategy in housing industry. Therefore, the development and application of energy-saving and insulated building materials have received extensive attention these years. With excellent properties of thermal insulation, acoustic insulation and light in weight, foamed concrete block has been one of the most promising materials these years [1]. Currently foamed concrete blocks on the market are mainly cement-based foamed concrete blocks [2-3], with high cost and large consumption of cement.

Fly ash is a solid waste produced from coal-fired power plants. The annual emission of fly ash in China is more than 200 million tons, but the utilization ratio is relatively low, which will seriously pollute the environment if not treated properly [4]. An important application of fly ash is to produce autoclaved aerated concrete block, which is similar with foamed concrete block in performance, with lime in the present of aluminum powder [5-7]. In order to reduce the consumption of lime and production cost, carbide slag (a kind of solid waste, rich in calcium hydroxide) may be used to substitute for lime [8-9]. However, it has negative effect on the aeration of the slurry and the initial strength of the aerated block is

* Corresponding author： zhaofq3366@126.com

low. Therefore the temperature in the slurry aerated block is the key problem in the use of carbide slag instead of lime.

For this reason, the fly ash slurry is mixed with foam directly, and a pre-curing process is used to increase the initial strength of the foamed concrete before cutting. GBFS and low-clinker cementitious material are used instead of OPC, to obtain high-performance lightweight block, achieving the synergistic effect among the materials and making the best use of various solid wastes.

2 Materials and methods

2.1 Materials

The fly ash used in the test is from Pingshan Power Plant; carbide slag is available for sale; cement used is the Ordinary Portland cement, strength grade 42.5; desulphurized gypsum is the industrial by-product of wet process desulfurization from coal power plant; GBFS is provided by a company in Tangshan, with a specific surface area of 408kg/m^2; low-clinker cement with the basic material ratio: GBFS 66.4%, steel slag 16.6%, cement 8%, gypsum 8%, sodium sulfate 1%; foaming agent, LG-2258 is bought from the market. The chemical compositions of the main raw materials are shown in Table 1.

Table 1. Chemical composition of main raw materials, wt%.

Raw material	SiO_2	MgO	CaO	Fe_2O_3	Al_2O_3	SO_3
Fly ash	54.41	0.63	7.01	5.84	27.15	0.81
Carbide slag	3.34	0.21	93.79	0.31	1.72	0.50
GBFS	39.05	8.47	30.98	13.65	2.53	-

2.2 Test methods

2.2.1 Preparation of foamed concrete block

(1) Mix fly ash, carbide slag, GBFS, gypsum, cement, low-clinker cement in proportion; stir well to form slurry A.
(2) Dilute foaming agent with water in a mass ratio of 1:60 and put in a high-speed mixer, foaming.
(3) Mix the stirred slurry A with the foam to prepare uniform foam contained slurry B.
(4) Pour the slurry B into 70.7 mm × 70.7 mm × 70.7 mm mold, pre-curing at 50℃ for 16 h.
(5) Cutting and stripping to form cubic samples.
(6) Put the test samples into the autoclave, curing at desired conditions. Current curing regime is "temperature rising 4h-180℃ constant temperature 4h-natural cooling".

2.2.2 Product performance test

The dry density and the compressive strength of foamed concrete of the samples are tested in accordance with JC/T 1062-2007. The samples were dried, weighed, and the average density of the samples is obtained with three samples. The mechanical strength values of the samples are tested on NYL-300A.

3 Results and discussion

3.1 Effect of GBFS content

The foamed concrete blocks were prepared based on the basic formula: fly ash 68.5%, carbide slag 20%, cement 10%, and 1.5% of gypsum. However, the initial strength values of the samples are too low. After autoclaving process, the final strength values of the samples are also very low. In order to increase the strength of the foamed concrete blocks, 10%, 15%, 20%, 25% and 30% of GBFS are added to the slurry respectively, the results are shown in Fig. 1.

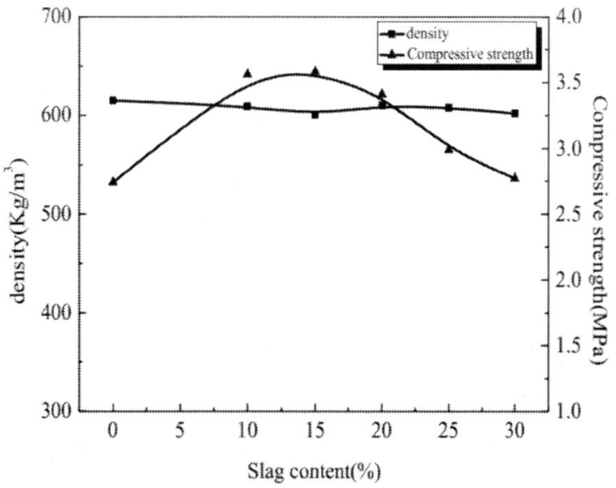

Fig. 1. Effect of GBFS content on performance of test block.

Fig.1 showed that the strength of the foamed concrete block is significantly increased after adding GBFS. With the increase of GBFS content, the compressive strength of the sample has a tendency to decrease. However, excessive GBFS content will bring more inert components, reducing the formation of the hydration products, and thus reduce the strength of the samples. Considering performance and cost, the optimal amount of GBFS is determined as 10%.

3.2 Effect of curing temperature

With the suitable GBFS content 10%, and the autoclaved curing regime "temperature rising 4h-constant temperature 4h-natural cooling", the performance of the test block is tested by changing the curing temperature. The results are shown in Fig. 2.

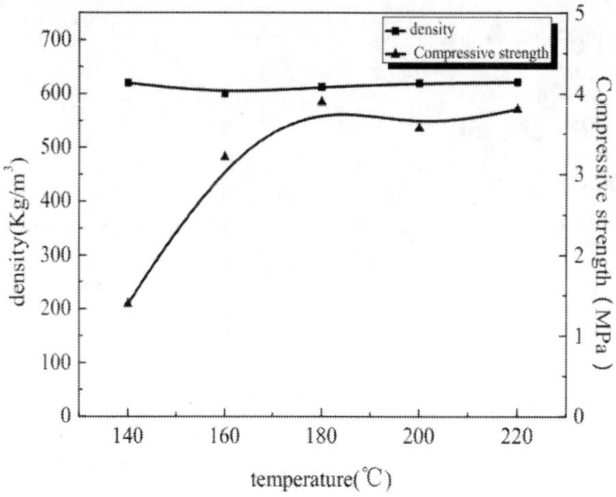

Fig. 2. Effect of curing temperature on performance of test block.

It shows that with the rise in curing temperature, the compressive strength of the block rises rapidly. When curing temperature is greater than 180℃, the strength of the samples tends to increase smoothly. With the increase in temperature, more active substances in the block react to form hydration products like hydrated calcium silicate, hydrated calcium aluminate and tombomite etc., improving the strength of the block. In summary, the appropriate autoclaved curing temperature is 180℃.

3.3 Effect of constant temperature time

With the mass ratio of fly ash, carbide slag, GBFS, cement and gypsum being 58.5: 20: 10: 10: 1.5, autoclaved samples were prepared under the curing regime "temperature rising 4h-180℃ constant temperature-natural cooling". The performances of the samples were investigated by changing constant temperature time. The results are shown in Fig. 3.

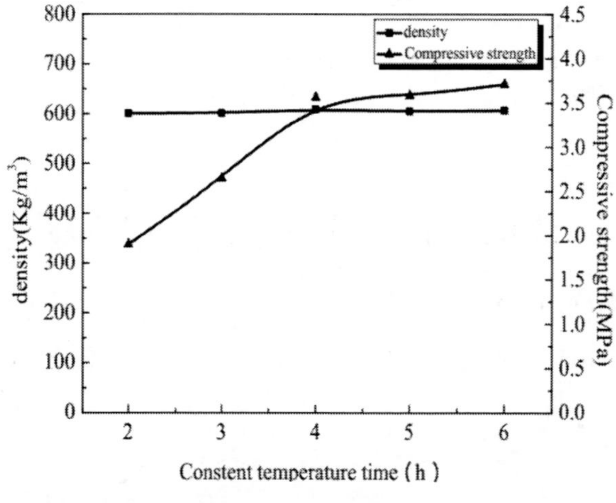

Fig. 3. Effect of constant temperature time on performance of test block.

It can be seen from Fig.3 that with the extension of the constant temperature time, the compressive strength of the foamed concrete block gradually increases. While constant temperature time is less than 4h, the compressive strength of the sample increases rapidly with the time. When the time is above 4h, the strength value tends to increase smoothly. The appropriate autoclaved constant temperature time is 4h.

3.4 Effect of cementitious materials

In order to further reduce the cost, save energy and reduce CO_2 emissions, low-clinker cement is used to substitute for OPC in 25%, 50%, 75%, 100%, respectively. The curing regime is "temperature rising 4h-180℃ constant temperature 4h-natural cooling". The results are shown in Fig. 4.

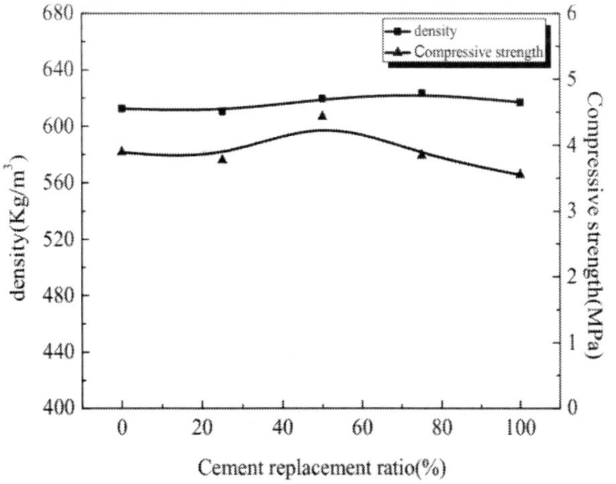

Fig. 4. Effect of low-clinker cement replacement on performance of test block.

As can be seen from Fig.4, with the increase in low-clinker cement replacement rate, the compressive strength of foamed concrete block increases first and then decreases, and all are greater than 3.5MPa. When the replacement rate is 50%, the strength reaches the peak. In order to reduce the cost, the optimal replacement ratio should be 100%, i.e. low-clinker cement can be used to substitute for OPC totally in the process.

4 Conclusions

(1) Using fly ash, carbide slag and low-clinker cement instead of cement, can produce foamed concrete with desired mechanical strength and density. This approach solved the problem of poor aeration when carbide slag is used to substitute for lime in autoclaved aerated block production, and make the best use of various solid wastes. It has the advantage of low cost, energy saving and friendly to the environment.
(2) It is necessary to use GBFS to substitute for fly ash partly. Adding 10% of the GBFS to the slurry in the production of foamed concrete block increases the initial strength of the block and shortens the pre-curing time.
(3) When GBFS content is 10%, curing regime is "temperature rising 4h-180℃constant temperature 4h-natural cooling", the cement can be replaced with low-clinker cement totally. The compressive strength of the block reaches 3.55 MPa while the density is 616.9kg/m^3. The autoclaved foamed concrete conforms to JC/T 1062-2007.
(4) Appropriate autoclaved curing regime is beneficial to the hydration process. Excessive curing temperature and too long constant temperature time contribute little to the block. The optimal autoclaved curing regime is "temperature rising 4h-180℃ constant temperature 4h-natural cooling".

References

1. WL Tian, QQ Xiong, FL Jiang. Performance Impact Factor Analysis of Fly Ash Foamed Concrete. J. Journal of Shijiazhuang Tiedao University (Natural Science), **26**(1):46-49(2013)
2. PT Gao. Discussion of foam concrete block property. J. Tile, 2:43-51(2012)
3. CQ Wang, KF Tan, XX Xu. Research status on foam concrete in China. J. Concrete, 12:57-62(2013)
4. S Zhang, GZ Li, WH Zhang, CC Jiang. Research on lightweight insulating concrete block with the composite cement fly ash foamer. J. Tile, 6:40-42(2011)
5. FQ Zhao, J Zhao , HJ Liu. Autoclaved brick from low-silicon tailings. J. Construction & Building Materials, **23**(1):538-541(2009)
6. FQ Zhao, B Guo, HJ Liu, & JQ Liu. Utilization of High Carbon Fly Ash and Coal Gangue in Blended Cement: A Case Study. International Conference on Bioinformatics and Biomedical Engineering (pp.1-4). IEEE.(2009)
7. PG Li, FQ Zhao. Autoclaved Brick from Steel Slag and Silicon Tailings. J. Advanced Materials Research, **878**:194-198(2014)
8. PG Li, FQ Zhao. Hydration intensification of autoclaved brick with high volume steel slag and tailings. J. Iron & Steel, **51**(10):84-90(2016)
9. SM Zhi, XX Jing,Z Wang. Research of Steam Curing Aerated Concrete by Substituting Lime for Carbide Slag. J. Construction Technology, **43**(24):105-107(2014)

Preparation of Magnesium Phosphate Cement and Application in Concrete Repair

Sainan Xing[1], and *Chengyou* Wu[1*]

[1]School of Civil Engineering, Qinghai University, 810016, Xining, China

Abstract. Potassium magnesium phosphate cement (MKPC) was prepared by using magnesium oxide (MgO) and potassium dihydrogen phosphate (KH2PO4, KDP). The effects of different calcination temperature, water-cement ratio and ratio on the operation of magnesium phosphate cement were studied. The effects of different water - cement ratio and ratio on the bonding strength of potassium magnesium phosphate cement and the application of potassium phosphate and magnesium cement in concrete repair and reinforcement were determined by experiments such as bending resistance, splitting, compression and bearing capacity. The experimental results show that the calcination temperature is 1100 ℃ , the calcination time is 1h, the settling time of magnesium oxide is 39min, and the compressive strength of one day reaches 53MPa. The water - cement ratio with the best bonding strength is 0.2, the mass ratio is 2:1.To meet the rapid repair of the construction requirements, is expected to be applied to the actual repair works.

1 Introduction

With the development of economy, people continue to improve the quality of life, and they have a higher requirement for travel quality. However, with the increasing number of road work, the load on the road is also growing. It means that appearance of pavement maintenance problems. Currently, the road in our country take cement concrete mainly, but the traditional maintenance method is generally the first to remove the damaged part, and then re-pouring, this method is not only time-consuming, and concrete to take longer maintenance time. At present, we know a kind of adhesive repair material is epoxy resin, but its functionalization research focus on experimental research. Due to its time-consuming, consumables, power consumption and type single and other disadvantages, so the test results are unpredictable. Meanwhile, epoxy resin achieve different functional applications often need to add functional filler, which hazardous to health and pollution of the environment. So it is necessary to study new materials that both to protect the environment, but also effective repair [1].

Magnesium phosphate cement prepared with lithium magnesium slag (MKPC) is a re-use of lithium waste and non-toxic, which is a new type of inorganic cementitious material with high bonding strength with old concrete. MKPC has the characteristics of early strength, fast hard, so that it can meet the requirements of rapid repair. MKPC saves

* Corresponding author: wuchengyou86@163.com

energy and protects environment as a new type of fast repair materials, can be widely used in roads, bridges, airport runways, which has attracted more and more attention from academics and the community[2]. This paper improves the maneuverability and volumetric stability of lithium magnesium slag magnesium phosphate cement by adjusting the calcination temperature and the way of adding fly ash to reduce the cost and obtain a new type of quick repair material which is economical, practical and environmentally friendly, and verify its feasibility in practical engineering.

2 Experiment

2.1 Raw materials.

The raw materials used in this experiment are by-product magnesium oxide (MgO) in the process of extracting Li_2CO_3 from Qinghai CITIC Guoan Technology Co., Ltd. and potassium dihydrogen phosphate (KH_2PO_4) produced by Hedong District Hongyan Reagent Factory in Tianjin. Table1 is the chemical composition of lithium magnesium slag content. Table 2 is the technical indicators of potassium dihydrogen phosphate.

Table 1. The composition of magnesium hydroxide residue after producing Li_2CO_3

Composition	Mg^{2+}	B_2O_3	Ca2^{+}	K^+	Na^+	Cl^-	SO_4^{2-}	Li^+	LOI*	Total
Mass fraction /%	59.22	4.91	0.33	0.027	0.36	0.92	0.45	0.25	32.55	99.02

Table 2. KH_2PO_4 technical indicators

Chemical composition	KH_2PO_4	SO4^{2-}	As	Fe
Percentage %	99.5	0.003	0.005	0.001

2.2 Experimental method.

The cement sample was manufactured in the case where KH_2PO_4 and MgO mass ratio and water-cement ratio unchanged. KH_2PO_4 and MgO in turn into the mortar, first dry mixing 2min, then added water at low speed stirring 1min, high-speed stirring 30s, and the stirred mixture was quickly poured into an iron mold of 10 mm x 10 mm x 10 mm, and shook it, and the specimen was placed at room temperature curing after demoulding. The strength of samples was tested at different ages[3]. The determination of the activity of magnesium oxide is shown in Ref4 [4].

3 Results and analysis

3.1 Effect of calcination temperature and time of lithium magnesium slag on MKPC.

As the magnesium oxide content is low and the condensation of magnesium phosphate cement coagulation is fast, it is necessary to carry out calcination treatment of lithium magnesium slag, the optimum temperature for calcination is determined by the activity of magnesium oxide and the compressive strength of the samples.

Table 3 shows that with the increase of calcination temperature and the prolongation of calcination time, the setting time and compressive strength of MKPC increases gradually.

But taking into account the nature of engineering applications, condensation time too fast causes the actual can't operate. Meanwhile, the initial setting time can't be too slow. To meet the requirements of engineering construction, choosing the calcination temperature (1100℃) and calcination time(1h).

Table 3. Compressive strength of MKPC specimen/MPa

Calcination temperature and time	1h	1d	3d	7d	Setting time / min
800℃ 3h	19.6	30.4	47.2	51.0	5
900℃ 1h	21.1	52.3	57.4	62.2	8
900℃ 2h	20.8	41.9	56.6	65.0	11
900℃ 3h	19.1	48.3	62.9	64.0	12
1000℃ 1h	21.0	51.9	62.5	68.8	15
1000℃ 2h	9.4	49.6	56.6	62.8	28
1000℃ 3h	-	33.5	42.5	52.2	63
1100℃ 1h	8.0	53.4	61.0	64.5	39
1100℃ 2h	-	39.1	50.2	52.9	65

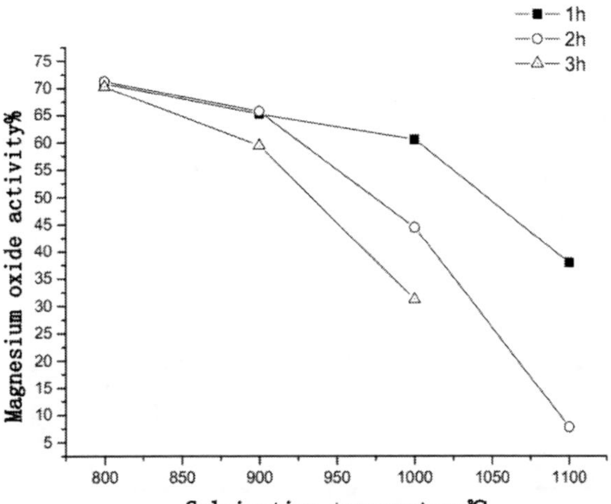

Fig.1. Relationship between the activity of magnesium oxide and the calcination temperature

Fig.1 displays that with the increase of calcination temperature, the activity of magnesium oxide decreases gradually in the case of the same calcination time. At the same calcination temperature, the activity of magnesium oxide decreases with the increase of calcination time. It means that the MKPC will enhance the operability in the later stage. It is necessary to further verify the accuracy of the effect of calcination temperature on the activity of magnesium oxide by staining test.

The principle of staining test [5] is acid –base neutralization. Take the citric acid solution 50g (11%concentration) in the beaker (including rotor), then add two drops of phenolphthalein indicator. Placing it in a discoloration tester, then add 15 g magnesium oxide powder. Discoloration time is from adding powder to discoloration of solution.

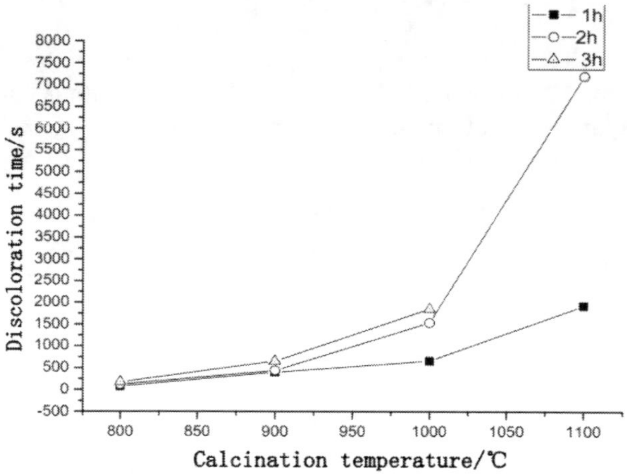

Fig. 2. Relationship between calcination temperature and discoloration time of MgO

Fig.2 implies that the activity of magnesium oxide decreases with increase of calcination temperature and calcination time. It also proves the accuracy of the test.

3.2 Influence of concrete strength ratio and water cement ratio on MKPC.

3.2.1 Plain concrete strength ratio.

When the water cement ratio was 0.2, the mass ratio of magnesium oxide and potassium dihydrogen phosphate is 1:1. The mass ratio of magnesium oxide and potassium dihydrogen phosphate was 1:1, 2:1, 2.5:1 and 3:1 respectively. The stirred mixture was quickly pasted onto the fracture surface of the concrete piece. Tested the flexural strength of 3h, 1d, 7d.

Table 4. Effect of MKPC mass ratio on bond strength/Mpa

(Temperature 1100℃ Insulation 1h W / C 0.2)

Mass ratio	3h	1d	7d
1:1	4.59	2.67	4.70
2:1	5.01	5.30	7.11
2.5:1	4.90	6.01	5.22
3:1	3.93	-	-

Table 4 demonstrates that In the case where the water-cement ratio is kept constant, the best mass ratio of the adhesive strength is 2:1, water-cement ratio of 0.2, the mass ratio of 2:1 magnesium phosphate cement (MKPC) paste the concrete rupture surface, the 7d flexural strength up to 7.11MPa, which indicates that MKPC has a strong paste repair performance.

3.2.2 Water cement ratio.

The mass ratio of magnesium oxide to potassium dihydrogen phosphate is 2:1, and the ratio of water to cement is respectively 0.16, 0.2, 0.25 and 0.3. The stirred mixture was quickly pasted onto the fracture surface of the concrete piece. Tested the flexural strength of 3h, 1d, 7d.

Table 5. Impact of water cement ratio on MKPC/Mpa

(Calcination temperature 1100℃ insulation 1h ratio 2:1)

W/C	3h	1d	7d
0.16	Too dry that on operate		
0.2	5.01	6.30	7.11
0.25	4.06	6.32	5.71
0.3	3.89	4.19	4.56

The experimental data shows that it is so dry that can't be pasted when the water-cement ratio is 0.16, the MKPC has the best paste strength while ratio of 2:1 and water-cement ratio of 0.2, because its flexural strength of 7d can reach 7.1MPa. When the ratio of water to cement is greater than 0.2, the flexural strength of MKPC bonded concrete specimen decreases with the increase of water-cement ratio, The reason for this phenomenon is that increase of water-cement ratio cause MKPC shrinkage, and lower bond strength[6].

3.3 Impact of splitting repair experiment on MKPC.

The split 100mm × 100mm × 100mm concrete test piece was bonded with water ratio of 0.2, mass ratio of 2:1 and fly ash parameters of 30% of the MKPC slurry. Placed at room temperature for curing, testing 3h, 1d, 7d splitting tensile strength [7]. The experimental results were 11.67KN, 18.51KN, 21.38KN respectively, compared with the original splitting tensile bearing capacity, the bearing capacity were restored to the original value of 36.53%, 57.93% and 66.92%. It is indicated that MKPC does have a strong ability to paste, and the bonding speed with the concrete quickly.

4 Conclusions

(1) There is an economical and environmentally friendly method that utilizing industrial waste residue to prepare magnesium phosphate cement. The production process of MKPC was studied. It was found by experiment that magnesium phosphate cement was prepared with lithium magnesium slag to have the calcination temperature of 1100℃.

(2) In order to apply it to the actual work of repairing concrete, the mass ratio of magnesium oxide to potassium dihydrogen phosphate should be 2:1, the water-cement ratio is 0.2.

(3) In the application of potassium phosphate and magnesium cement engineering, it is found that potassium phosphate magnesium cement has a strong bonding capacity, fast hard, early strength through the bending, split and the others is a good choice for quick repair works.

(4) The preparation of magnesium phosphate cement in lithium magnesium slag, is a kind of environmentally friendly and economical magnesium phosphate cement preparation method, although there are still many problems can't be resolved, it will be more research and exploration on this project. In the future life of lithium magnesium slag prepared by potassium magnesium phosphate cement (new inorganic cementitious materials) can be widely used in roads, bridges, airport runways and other repair works.

This study was supported by the National Natural Science Foundations of China (grants No.51662035,

No.U1407104, No.51678304 and No.21401209), Natural Science Foundations of Qinghai Province (grants No.2015-ZJ-947Q and 2015-ZJ-937Q) and foundation teaching and research innovation team project of higher education institution of Qinghai province.

References

1. Z Ruan, Z H Liu, Z P Deng, S Y Cheng, S T Ye. Research Progress in Functionalization of Epoxy Resin Systems. J. **12**,8 (2015)
2. L R Wu. Study on Modification of Magnesium Phosphate Cement by Fly Ash and Its Engineering Application. J. (2014).
3. Y S Tan, H F Yu, Y Li, J M Dong, et al. Preparation of Fly Ash Magnesium Phosphate Cement Using Byproduct Magnesium Oxide Containing Boron from Salt Lake. J. **42**, 8 (2014).
4. J M Dong, H F Yu, L M Zhang. Study on Experimental Conditions of Hydration Methods of Determining Active Magnesium Oxide Content. J. **18** , 4 (2010).
5. J P Yuan, C Wang, K Y Shen, D L Xiao. Commentary on Anti Tarnishing Experiment and Evaluation Method of Silver Alloy. J. **8**, 4 (2006).
6. Q B Yang, B Y Lei. Effect of Fly Ash on Adhesive Properties of Phosphate Cement Mortar and Concrete. J. 3 (2016).
7. W H Chen, S S Cui. Flexural Behavior of Prestressed Concrete Beams Strengthened with CFRP Sheets Bonded with Inorganic Adhesive. J. **36**, 5(2014).

Composite Design of Low Thermal Conductivity Mullite Brick for Application to Cement Kiln

Songlin Chen [1, 3], *Juntao* Wang [1, b], *Lin* Yuan [1, 2 *], and *Quanyou* Li [2]

[1] Ruitai Materials Technology Co., Ltd., China Building Materials Academy, Beijing China, 100024
[2] Zhengzhou Ruitai Materials Technology Co., Ltd, Zhengzhou Henan China, 452374
[3] Yixing Refractories Co., Ltd., Yixing Jiangsu China, 214264

Abstract. In order to restrain cement kiln shell's thermal loss and potential safety hazard, composite design is employed of working layer, thermal-preservation layer and thermal insulation layer. Thermal-preservation layer's boundary and strain is analyzed and overall thermal conductivity is calculated. The result shows that sinusoidal curve boundary combination between working layer and thermal-preservation layer, slope-shape with angle of 135 degrees between thermal-preservation layer and thermal insulation layer could relieve strain concentration. Composite mullite brick comprehensive thermal conductivity decreases from the working layer $2.74 W \cdot m^{-1} \cdot K^{-1}$ to $1.50 W \cdot m^{-1} \cdot K^{-1}$. When applied in cement kiln, temperature of kiln shell is lower 50℃ to 70℃ than using normal bricks.

1 Introduction

Cement rotary kiln is high energy consumption mechanical equipment. Especially in the latter and upper transition zone, refractory bricks directly contact with clinker without the preservation of coating. Higher thermal conductivity leads to high temperature of kiln shell, high heat loss and high fuel consumed, lower lifetime of tire and kiln shell. Meanwhile, under the impact of chemical corrosion and mechanical erosion, refractory is easily damaged [1, 2].

In order to restrain kiln shell's thermal loss and potential safety hazard, method of lower conductivity coefficient of refractory will be the first alternative. But lower conductivity coefficient method of SiC-mullite brick and Magnesia-spinel brick is limited. When choosing other materials, corrosion resistance property and environment issues should be taken into consideration. Given all that, composite designs are of considerable interest [3-6]. Multilayer mullite based brick with reduced thermal conductivity choose the method of multilayer structure with working layer, thermal-preservation layer and thermal-insulation layer. Working layer has been designed with high strength, lower thermal conductivity, high refractoriness under load, high thermal shock resistance and long lifetime; As bonding structure connecting working layer and thermal-insulation layer, thermal preservation layer

* Corresponding author: yl@bjruitai.com

combines excellent high strength as well as lower thermal conductivity; thermal insulation layer use ultra-lower thermal conductivity to decrease brick's thermal conductivity.

2 Composite design of multilayer mullite based brick with low thermal conductivity

2.1 Refractory in working layer

Working layer choose andalusite-SiC-mullite brick, which is based on M70 bauxite based mullite, M70, M80 homogeneous materials and andalusite. After mixing, forming and 1490 ℃ fired, main indexes of working layer could satisfy basic requirements of transition zone ($Al_2O_3 \geq 65\%$, apparent porosity≤20%, Bulk Density 2.65g/cm^3, Cold Crushing Strength≥90MPa, Refractoriness under load $T_{0.6} \geq 1620$ ℃). The thermal conductivity coefficient of this area is 2.74W·m^{-1}·K^{-1}.

2.2 Refractory in thermal-preservation layer.

Aiming for lowing overall thermal conductivity, this layer should share lower conductivity coefficient than working layers, have adhesive ability with working layer, as well as excellent cold modulus of rupture and abrasion resistance. This layer is based on M60 bauxite based mullite, clay, Al_2O_3 powder and silicon micro-powder. Then high-temperature fired contributes to the formation of mullite phase, which guaranteed high strength and similar expansion coefficient and linear change on reheating with working layers. Then formed mullite phases obvious reduced thermal conductivity, and micro-powders could produce closed pores in the intern structure, which further decrease thermal conductivity [7, 8]. The thermal conductivity of this layer is 1.58W·m^{-1}·K^{-1}.

2.3 Refractory in thermal-Insulation layer

The ultimate quality of multilayer mullite based brick with reduced thermal conductivity is directly influenced by the quality of thermal-insulation layers. This layer is composed of zirconia ceramic fibers of high refractoriness and working accuracy, the thermal conductivity coefficient is 0.12W·m^{-1}·K^{-1}.

2.4 Thickness and shape design of multilayer brick.

2.4.1 Thickness design of different layers

In order to ensure lifetime and operation rate of cement kiln, when residue thickness is less than 1/3 after erosion and abrasion, kiln must be stopped to maintenance. Therefore, the thickness of working layer is 70% of total thickness. The thickness of working layer is 0.140m, thermal-preservation layer is 0.035m, and thermal-insulation layer is 0.025m.

2.4.2 Shape-design of thermal preservation layer

The preservation layer is located between working layer and thermal insulation layer, which two faces of preservation layer connect with the other two layers. Therefore, thermal strain concentration should be taken into consideration.

2.4.2.1 Boundary-shape design between thermal-insulation layer and thermal-preservation layer

The heat preservation performance of multilayer bricks is not only related to thickness of materials used in thermal insulation layer and thermal preservation layer, but also related to the area of materials. Furthermore, the thermal preservation layer has capability of burdening load. Therefore, the thickness of thermal-insulation layer is proportional to heat preservation layer, but weaker the brace parts. Numerous trials have shown that open-shape area of brace, as shown in Fig.1, is one third of overall area, extending to inside at the angle of 135°.

Fig.1. Open shape of thermal preservation layer

2.4.2.2. Boundary shape design

Materials in thermal-preservation layer and working layer have similar composition, physical properties. This leads to excellent binding, cold modulus of rupture and abrasion-resistance. Therefore, boundary is designed sinusoidal curve boundary combination, as shown in Fig.2.

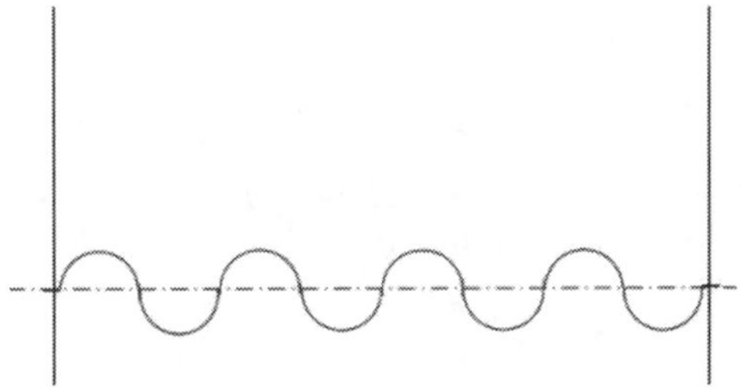

Fig.2. Sinusoidal curve boundary of working layer and thermal-preservation layer

Therefore physical design of multilayer brick is shown in Fig3. Boundary of working layer and thermal-preservation layer is sinusoidal curve design, with the angle of 135°.

Fig.3. Multilayer mullite based brick with reduced thermal conductivity

2.4.2.3. Strain analysis of structure in thermal- preservation layer

When thermal-preservation layer is influenced by vertical strain caused by gravity of clinker, strain from central part and two edges could be dissipated along angle of 45°, as shown in Fig4. The central part will not be damaged by strain.

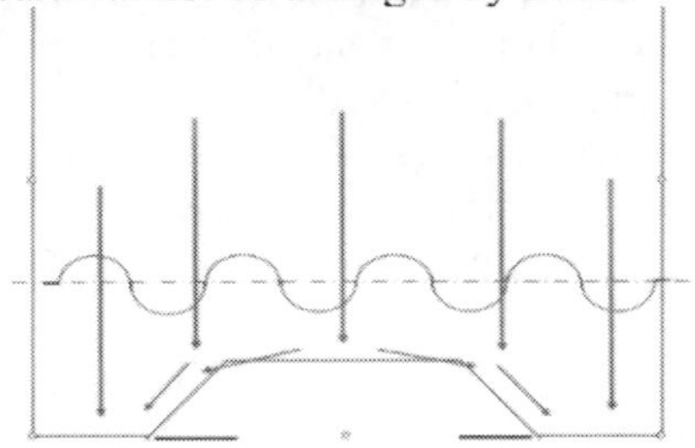

Fig.4. Vertical strain on thermal-preservation layer

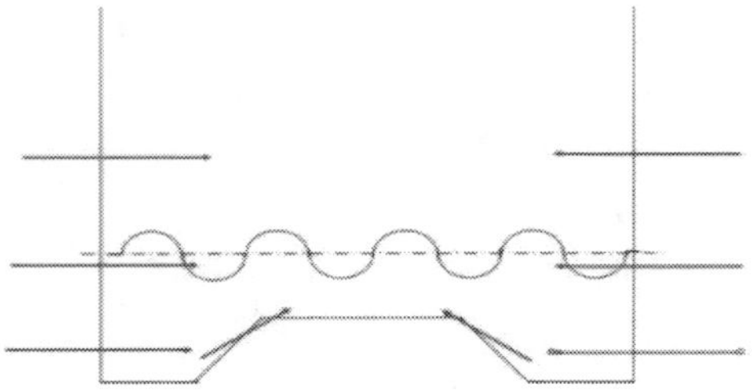

Fig.5. Transverse expansion stress analysis

When thermal-preservation layer is squeezed by transverse expansion stress between bricks, the pressure of foothold is diffused along 45°, as shown in Fig.5, reduces the rates of foothold parts' fracture.

When circumferential shear loads impact on thermal-preservation layer, angle of 45° contribute to increased contact area 50% (as shown in Fig.6), which decrease the possibility of crack caused by circumferential shear loads on brace parts.

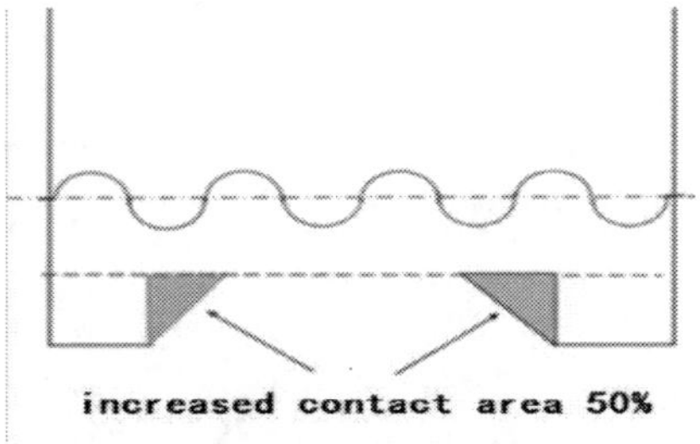

Fig.6. Circumferential shear loads analysis

3. Comprehensive thermal conductivity coefficient calculations

According to heat transfer calculation formula,

$$q = \frac{\lambda}{\delta} \cdot \Delta T \tag{1}$$

Under the circumstance of multilayer,

$$q = \frac{\Delta T}{\sum_{i=1}^{n} \frac{\delta_i}{\lambda_i}} \tag{2}$$

In formula (1) and (2), q is quantity of heat, δ is thickness of different layers, λ is thermal conductivity of different layers, ΔT is temperature difference. Considering formula (1) and (2), thermal conductivity of multilayer brick can be described as formula (3),

$$\lambda_{\text{compound}} = \frac{\delta}{\sum_{i=1}^{n} \frac{\delta_i}{\lambda_i}} \tag{3}$$

For the thickness of first layer is 0.14m, thermal conductivity is $2.74W \cdot m^{-1} \cdot K^{-1}$; thickness of second layer is 0.035m, thermal conductivity is $1.58W \cdot m^{-1} \cdot K^{-1}$; while thickness of third layer is 0.025m, thermal conductivity coefficient is $0.113W \cdot m^{-1} \cdot K^{-1}$. Therefore the overall thermal conductivity is $1.50W \cdot m^{-1} \cdot K^{-1}$.

When multilayer bricks applied in cement kiln system, the situation is quite like cylinder thermal loss. If the radius of kiln system is d_1=2.4m, while the first layer radius is d_2=2.54m, the second layer radius d_3=2.575m, the third layer radius d4=2.6m. Assumed that temperature of kiln system is T_1=1300℃, room temperature T_2=25℃。

Assumed that kiln shell temperature is T_3, so heat dissipating capacity can be calculated as formula (4),

$$q_{\text{inner}} = \frac{2\pi \sum_{i=1}^{n} (t_n - t_{n+1})}{\sum_{i=1}^{n} \frac{1}{\lambda_n} \ln \frac{r_{n+1}}{r_n}} = \frac{2\pi (T_{\text{inner}} - T_{\text{shell}})}{} \tag{4}$$

Heat dissipating capacity of kiln shell is

$$q_{\text{external environment}} = a(T_{\text{shell}} - T_{\text{external environment}})\pi d_4 \tag{5}$$

When heat dissipating capacity reach balanced situation,

$$q_{inner} = q_{external\ environment} \tag{6}$$

Presumed that coefficient of convective heat transfer is $38W \cdot m^{-1} \cdot K^{-1}$. After calculation, Temperature T_{shell} is 290℃, lower about 60℃ than normal kiln shell temperature (350℃).

4. Applied to cement application

In July 2013, Multilayer mullite based brick with reduced thermal conductivity used in area of 37 to 47 meters in cement rotary kiln of Dequan Jilin. Temperature of kiln shell is 200℃ after using one week; while the temperature is after 220℃ using one month, the temperature is 230℃ after using two months. If normal bricks used in cement furnace, the average temperature is 370℃. As conclusion Multilayer brick could lower kiln shell temperature more than 100℃, substantially abate thermal loss and decrease the possibility about elevated temperature of kiln shell. Due to thermal preservation effects after several months' usage condition, widespread use leads to better thermal preservation effects.

5. Conclusions

1) Sinusoidal boundary and 135° shape design, which avoid thermal preservation layer's strain concentration, and avoid broken of supporting legs.

2) Composite design of working layer, thermal preservation layer and thermal insulation layer is employed to lower thermal conductivity ($2.74W \cdot m^{-1} \cdot K^{-1}$ to $1.50W \cdot m^{-1} \cdot K^{-1}$);

3) When multilayer brick has been chosen in transitional zone of cement furnace, the kiln shell temperature is lower about 60℃ (rang from 50℃ to 70℃).

The authors would like to thank the financial support from the National Key Technology Research and Development Program of China (No.2013BAE03B01-10) and Yixing Key Research and Development Plan (No.YGB2016020).

References

1. Q. S. Zhou. Preparation of SiC/mullite/alumina composite ceramic. D. Central South University: (2001).(In Chinese)
2. F. Yao, Z. Y. Liu. China Non-metallic Mining Industry. J. 6, **8** (2013) (In Chinese)
3. L, Yuan, X. F. Chen. Naihuo Cailiao. J. 50, **161** (2016). (In Chinese)
4. Z. Y. Liu, B. Xu. J. of Materials Science & Engineer. J. 32, **85** (2014). (In Chinese)
5. J. Z. Wang, L. Yuan. Naihuo Cailiao. J48, **161** (2014). (In Chinese)
6. S. J. Chen Research on thermo mechanical stress of ladle composite structure and its life prediction. D. Wuhan: Wuhan University of Science and Technology, 2007. (In Chinese)
7. J. F. Chen, H. Z. Zhao. Naihuo Cailiao. J. 49, **120** (2015). (In Chinese)
8. X. Hou, Q. H. Xue.. Bulletin of the Chinese ceramic society. J. 34, **2325** (2015). (In Chinese)

Study on Hot Deformation Behavior of High Strength Construction Steel 34CrNiMo6

Feng Kang [1]*, *Echuan* Yang[2], *Jun* Lin [1], and *Chuankai* Hu [1]
[1]Southwest Technology and Engineering Research Institute, Chongqing 400039, China
[2]Chongqing University of technology, Chongqing 400054, China

Abstract. The hot pressure deformation tests were carried out by using Gleeble-2000 simulator at the temperature of 950 ℃ ～ 1150 ℃ and strain rate of $0.001s^{-1}$ ～ $1s^{-1}$ for a kind of high strength construction steel 34CrNiMo6 for Crankshafts and the curve of its flow stress were obtained. By studied the data, the flow stress constitutive equation, hot deforming active energy and the Z parameter were obtained. The curve of flow stress of 34CrNiMo6 steel indicated that the flow stress was increased with the decrease of temperature or increase of the strain rate. During the deformation, temperature and strain rate take important effect on dynamic recrystallization and dynamic recovery. It was propitious to dynamic recrystallization with raising temperature or reducing strain rate, and also made for the refinement of grain.

1 Introduction

34CrNiMo6 steel is the high-strength structural steel. It is the material meeting the German standards, and it is the hardened and tempered steel. It has good comprehensive mechanical properties, which can be widely used in manufacturing the bearing parts with large section and complex shape, such as the bent axle of motor, connecting rod and cam etc. A factory now uses such material to manufacture the bent axle of motor after all-fiber forging. As its thermoplastic deformation behavior affects the product quality (usability and performance) directly, thus it is very important to decide the reasonable thermal deformation parameters for improving the material organization and performance. However, the existing research on such material mainly focuses on the thermal deformation process parameters and heat treatment process etc, the report about researched on thermal deformation behaviors of such material is rare[1-3], therefore, in this thesis, the hot compression simulation test is adopted in the research of thermal deformation behaviors of 34CrNiMo6 steel so as to provide the experimental theory for developing plastic processing technology and improving the material organization performance.

* Corresponding author: tiannakaff@163.com

2 Test methods

The 34CrNiMo6 material for test is taken from the Φ150mm annealed forging material; the chemical composition is shown in Tab.1. The diameter of the sample is Φ10mm.

Table 1. Chemical composition of 34CrNiMo6 steel in wt%

C	Cr	Ni	Mo	Mn	Si	P	S
0.37	1.44	1.37	0.25	0.66	0.26	0.015	0.012

The test is conducted on the thermal simulated test machine of Gleeble-2000 type. The argon is adopted as the protective gas to reduce the oxidation in deformation, and the graphite is filled as the lubricant in both ends of the sample to achieve the nearly uniform deformation of the sample. At first, the sample is heated to 1200℃ with heat preservation of 5min, and then cooled to the test temperature in the speed of 5℃/s with heat preservation of 20s, and the temperature gradient generated in the temperature changing process is eliminated. The deformation temperatures of this compression test shall be: 950℃, 1000℃, 1050℃, 1100℃ and 1150℃; the deformation rates shall be: $1S^{-1}$, 0.1 S^{-1}, 0.01 S^{-1} and 0.001 S^{-1}. The maximum true strain is 0.9.

3 Stress-strain analyses

Fig. 1 is the stress-strain curve under the test conditions of 34CrNiMo6 steel. According to the figure, the flow stress is increased with the decrease of deformation temperature and the increase of strain rate. Because under the same strain rate, when the temperature is increased, the dynamic softening speed is increased, and if the softening degree is enlarged, the peak stress and steady flow stress shall be decreased; however, when the deformation temperatures are the same and the strain rate is increased, then the rate of work hardening shall be increased as well, and the peak stress and steady flow stress shall be increased correspondingly. When the strain rate is low, obvious peak occurs in the curve (Fig.2a,b), which is presented as dynamic recrystallization type curve; and when the strain rate is $0.1S^{-1}$, the peak value (Fig. 2c) only occurs in the curve when the temperature increases to 1050℃ or above, and dynamic recrystallization occurs. When the temperature is lower than 1050℃, the deformation process only has dynamic response; when the strain rate is increased to $1S^{-1}$, no obvious peak value in curve under all temperatures (Fig. 2d), and only dynamic response occurs in the deformation process, which is presented as dynamic response curve. Therefore, in the deformation process of such material, the higher the temperature is, the lower the strain rate shall be, and it is good for dynamic recrystallization, so that the grain refinement of material, organization and performance can be improved.

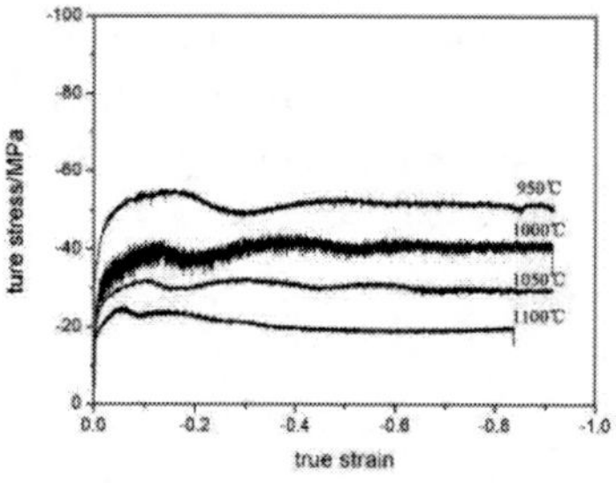

a. $\dot{\varepsilon} = 0.001S^{-1}$

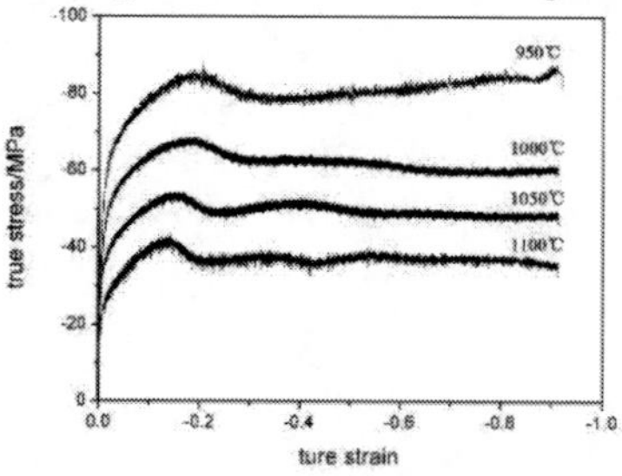

b. $\dot{\varepsilon} = 0.01S^{-1}$

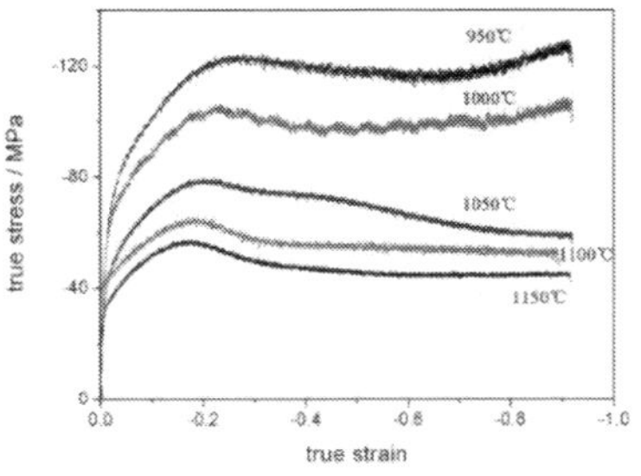

c. $\dot{\varepsilon} = 0.1S^{-1}$

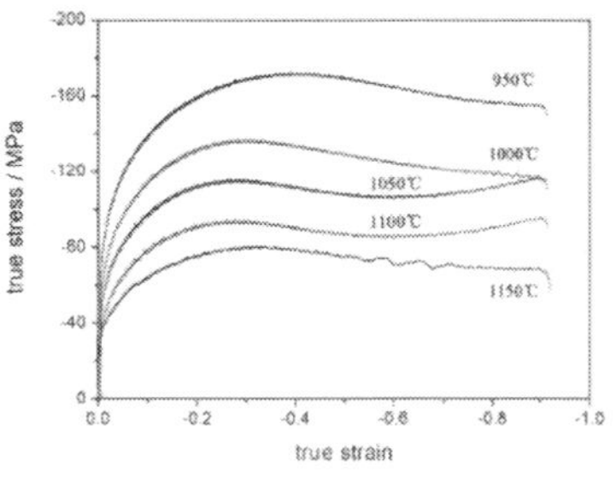

d. $\dot{\varepsilon} = 1S^{-1}$

Fig. 1. The stress- strain curves of 34CrNiMo6 steel

According to relevant researches, the flow stress of the material is not only connected to the chemical components, but also related to parameters such as strain rate and deformation etc. According to researches on different hot working parameters, the following function relationship[4-5] exists between the flow stress and the deformation conditions:

$$\dot{\varepsilon} = A[\sinh(\alpha\sigma)]^n \exp(-\frac{Q}{RT}) \tag{1}$$

Where, $\dot{\varepsilon}$ is the deformation rate, σ is the peak stress or steady-state flow stress (the peak stress in this paper), Q is the deformation activation energy (reflect the complexity of thermal deformation), A, R and n are the constants, and T is the deformation temperature.

Where, when the stress is low ($\alpha\sigma \leq 0.8$), sinh(ασ)n in the equation (1) can be simplified into σn. When the stress is high ($\alpha\sigma > 1.2$), sinh(ασ)n in the equation (1) can be simplified into exp(βσ); where $\alpha = \beta/n$, n and β are constants[6,7].

During the high-temperature deformation process of the metal, to indicate the influence of deformation temperature and stress rate on the thermal deformation process, the Zener-Hollomon parameter[8] is introduced (the stress rate factor with temperature compensation: Z parameter):

$$Z = \dot{\varepsilon}\exp(Q/RT) \tag{2}$$

Substitute two simplified methods in equation (1) into (1), we can know that:

$$\dot{\varepsilon} = A\sigma^n \exp(-Q/RT) \tag{3}$$

$$\dot{\varepsilon} = A\exp(\beta\sigma)\exp(-Q/RT) \tag{4}$$

Take the logarithm in both sides of equation (3) and (4):

$$\ln\dot{\varepsilon} = \ln A + n\ln\sigma - Q/RT \tag{5}$$

$$\ln\dot{\varepsilon} = \ln A + \beta\sigma - Q/RT \tag{6}$$

Under given temperature: seek the partial derivative on $\ln\dot{\varepsilon}$ in (5) and (6) separately, then:

$$n = [\frac{\partial(\ln\dot{\varepsilon})}{\partial\ln\sigma}]_T \tag{7}$$

$$\beta = [\frac{\partial(\ln\dot{\varepsilon})}{\partial\sigma}]_T \tag{8}$$

Plotting the data obtained from thermal simulation compression test according to equation (7) and (8), which is as shown in Fig.2:

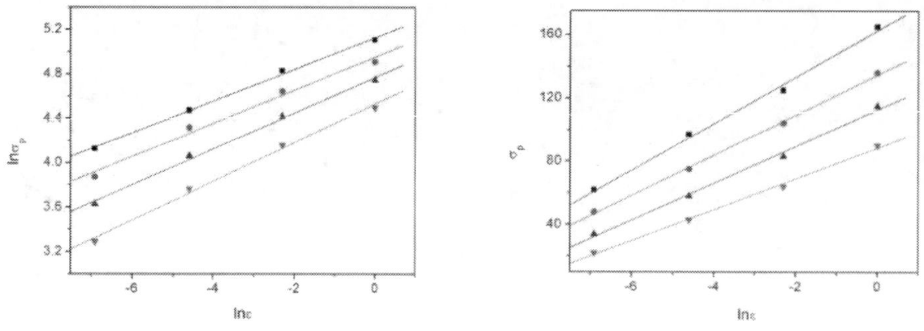

a. $\ln\sigma_p - \ln\dot{\varepsilon}$ b. $\sigma_p - \ln\dot{\varepsilon}$

Fig.2. Relationship between peak stress and strain rate

According to the figure, both $\ln\sigma_p - \ln\dot{\varepsilon}$ and $\sigma_p - \ln\dot{\varepsilon}$ indicate obvious and approximate linear relationship, and implement the regression analysis with least square method; it gets n=6.36102, β=0.08202 and α=β/n=0.012894.

Take the logarithm in both sides of equation (1):

$$\ln\dot{\varepsilon} = \ln A + n\ln\sinh(\alpha\sigma) - Q/RT \tag{9}$$

When $\dot{\varepsilon}$ is constant, seek the partial derivative on 1/T in equation (5), we can know that:

$$Q = Rn\left[\frac{\partial\ln\sinh(\alpha\sigma)}{\partial(1/T)}\right]_{\dot{\varepsilon}} \tag{10}$$

When T is constant, seek the partial derivative on $\dot{\varepsilon}$ in equation (9), we can know that:

$$1/n = \left[\frac{\partial\ln\sinh(\alpha\sigma)}{\partial\ln\dot{\varepsilon}}\right]_T \tag{11}$$

According to equation (10) and (11), $\ln\sinh(\alpha\sigma) - \ln\dot{\varepsilon}$ and $\ln\sinh(\alpha\sigma) - 1/T$ can be all in linear relation; relation curve made according to test data is shown in Fig. 3 and Fig. 4:

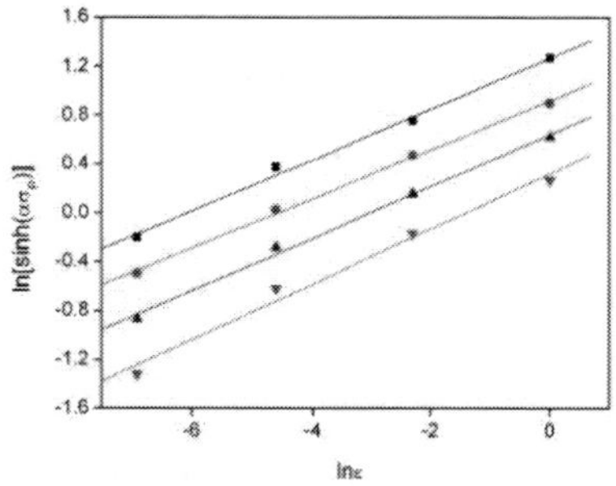

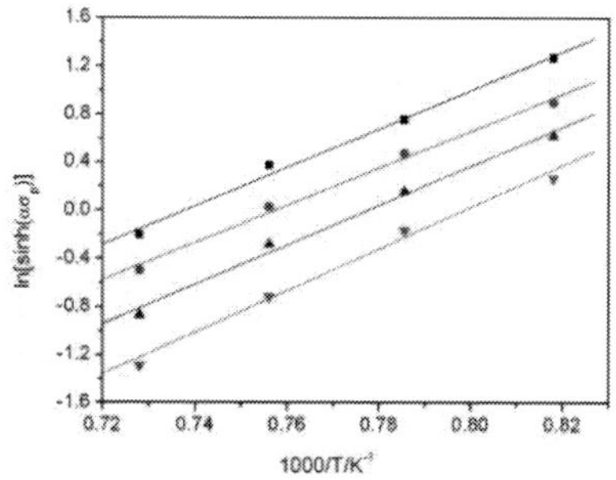

Fig. 3. Relationship between $\ln\sinh(\alpha\sigma)$ and $\ln\dot{\varepsilon}$ **Fig. 4.** Relationship between $\ln\sinh(\alpha\sigma)$ and 1/T

Carry out linear regression analysis of Fig.3 and Fig.4; the obtained activation energy Q of hot deformation and other constant data are shown in Table 2:

Table 2. Parameters in hot deforming formulation

Q, kJ/mol	n	α, MPa	R, $J\cdot mol^{-1}K^{-1}$
636.936	4.7102	0.012894	8.31

According to data in Table 2, parameter Z can be expressed as: $Z = \dot{\varepsilon}\exp(636936/8.31T)$

According to equation (1) and (2), there is:

$$Z = A[\sinh(\alpha\sigma)]^n \quad (12)$$

Take the logarithm of both sides, and get:

$$\ln Z = \ln A + n\ln[\sinh(\alpha\sigma)] \quad (13)$$

According to equation (13), $\ln Z$ and $\ln[\sinh(\alpha\sigma)]$ are in linear relation; Fig. 5 which is made based on test data is:

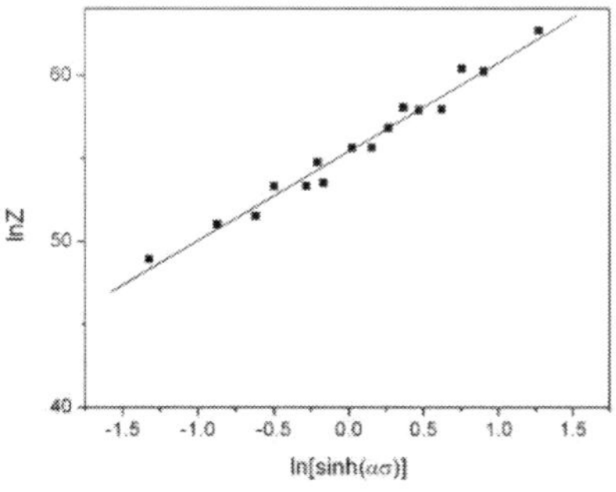

Fig. 5. Relationship between $\ln\sinh(\alpha\sigma)$ and $\ln Z$

From Fig.5 we can see, all data got from the test are near the regression straight line, which shows that the linear correlation is in high degree; the test data well reflected equation (12) that with the increase of deformation temperature and the decrease of strain rate, the parameter value of Z is decreasing, the peak stress is also decreasing, while dynamic recrystallization is also liable to come up. Calculate the intercept of this regression straight line, and lnA can be got, while constant A=2.35×1016 in equation (13) can further be determined.

Take all data into equation (1), and get the flow stress constitutive equation of the steel:

$$\dot{\varepsilon} = 2.35\times10^{16}[\sinh(0.012\sigma)]^{4.71}\exp(-\frac{636936}{RT})$$

4 Summaries

(1) Hot deformation activation energy Q of 34CrNiMo6 high-strength structural steel for this bent axle is 636.936 kJ/mol, and its hot deformation equation is $\dot{\varepsilon} = 2.35\times10^{16}[\sinh(0.012\sigma)]^{4.71}\exp(-\frac{636936}{RT})$.

(2) Hot deformation activation energy Q of 34CrNiMo6 steel is 636.936 kJ/mol, while $Z = \dot{\varepsilon}\exp(636936/8.31T)$.

(3) During the deformation process, both the deformation temperature and strain rate play an important role in dynamic recrystallization and dynamic recovery of 34CrNiMo6 steel; while whether the increase of deformation temperature or decrease of strain rate is good for the appearance of dynamic recrystallization and grain refinement of deformed materials in the deformation process.

References

1. J.L. Zhang, W.H. Zang, R.D Fu. et al. Hot deforming behaviors and microstructures of 2.25Cr-1Mo-0.25V steels[J]. Journal of plasticity engineering, 2010, **3** (17):44-48
2. B. Zhang, H.B. Zhang. Hot deformation behavior of 35CrMo steel[J]. Acta. Metallurgica Sinica, 2004, **40**(10):1109-1114
3. M. Wang, H.P. Hong, X.M. Wang. Thermal deformation behavior and flow stress model of 37Mn5 steel[J]. Journal of Wuhan University of Science and Technology, 2010, **33**(2):134-137
4. B. J. Zhang, G.P. Zhao, L.Y. Jiao. et al. Influence of hot working process on microstructures of superalloy GH4586[J]. Acta. Metallurgica Sinica, 2005, **41**(4):351-356
5. P.F. Liu, D. Liu, Z.J. Luo. et al. Flow behavior and dynamic recrystallization model for GH761 superalloy during hot deformation[J]. Rare Metal Materials and Engineering, 2009, **28**(2):275-280
6. Jonas J J, Quelennec X, Jiang L. The Avrami kinetics of dynamic recrystallization[J].Acta materialia, 2009,**57**:2748-2756
7. C.X. Wang, X.M. Liu, Z.L. Tian. et al. Hot Deformation Behavior of 23Co14Ni12Cr3MoE Ultra-High Strength Steel[J]. Journal of Aeronautical Materials, 2011, **31**(6):19-22
8. Almagro J F, Llovet X, Heredia M A, et al. Soluble fraction of stabilizing elements in ferritic stainless steel[J]. Microchim Acta, 2008, **161**(3/4):323-327

A Calculation Theory of Earth Pressure Considering Microstructure of Porous Materials

Jun-hu Li [1*], *Chao* Zhang [2], *Wei* Xue [3,4], *Wen-chao* Zhang [3,4], and *Ri-qing* Xu [1]

[1] Zhejiang University, Research Center of Coastal and Urban Geotechnical Engineering, 310058 Hangzhou, China
[2]China Road and Bridge Corporation, 100011 Beijing, China
[3]Guangzhou Chemical Grouting Engineering Co., Ltd., CAS, 510650 Guangzhou, China
[4]Guangdong Province Chemical Grouting Engineering and Materials Academician Workstation, CAS, 510650 Guangzhou, China

Abstract. To explore a more reasonable earth pressure calculation theory, microstructure characteristics of soft clay in Hangzhou, Ningbo and Wenzhou had been observed using scanning electron microscope (SEM) and were quantitatively analysed using Image-Pro Plus (IPP) software. The relationship between porosity and contact area of soft clay was got by considering the microstructure. Combining soil lateral stress transfer principle, a calculation theory of earth pressure considering soil contact area was got. The possible reason of the differences between earth pressure and the actual monitoring earth pressure was analysis by the case.

1 Introduction

The calculation of earth pressure is one of the primary research topics of soil mechanics. The method of estimating water and earth pressures together or separately can't simply meet the needs of engineering, while these two methods are still under discussion. With the development of urban construction and the exploitation of the underground space, especially the construction of large underground emporium and subway tunnels, there is growing concern about the working condition of different kinds of supporting structure during the design and construction. Anecdotal evidence suggests that the interaction between the soil and the supporting structures is much more difficult than what we have learnt, and the method to calculate the earth pressure on the back the soil retaining structures is still under discussion.

As for these questions, the scholars domestic and abroad have carried out plenty of qualitative and quantitative research work on microstructure of fine particles based on microscope images in the past few decades [1-5]. Chen Yu-jiong [6] believed that, as far as the method of estimating water and earth pressures together manages to avoid the puzzle of the pore water pressure by seeing the water and soil as an integral using total stress strength parameters, there is no need to figure out the mechanical mechanism of the pore water pressure.

* Corresponding author: 285014418@qq.com

But Li Guang-xin [7] pointed out the insufficiency of this method and believed that this method makes sense from microcosmic point of view. Besides, he proposed to incorporate the view of microstructure into the study of stress transmission mechanism. On the basis of previous studies, the paper analyses soft clay microstructure and modifies the commonly used formula of earth pressure, followed by a calculation theory considering soil contact area.

2 The relationship between soil contact area and porosity

2.1 The micro porosity of the clay

Fig.1. is the microscopic pore figure of some kind of soft clay observed using scanning electron microscope (SEM). We can see from this figure that the gray level of the pore of soft clay decreases from deep to shallow. In the bottom of the pore, the gray level is highest and the color is darkest, which is opposite on the opening of the pore. The sample used in experiments shows the section area is getting bigger from the bottom to opening.

Fig.1. Microscopic pore figure of soft clay

Fig.2. Three-dimensional display of particle surface

In the SEM figure, the pixel represents the area of the figure and the gray level stands for the colour depth. Different gray levels can be expressed by different thresholds, of which the values range from 0 to 255. We set the brightness of the deepest place as the minimum value, and when the threshold goes larger, the brightness varies from deep to shallow.

Wang Bao-jun [8], basing on the gray level, with the help of ArcGIS software, turns the SEM figure into the three-dimensional display image of particle surface which shows the 3D distribution of the pore in the soil. In the three-dimensional space model based on this idea, which is shown in fig.3., the area of plan in the horizontal direction can be expressed by the pixels, while the vertical height can be characterized by the threshold.

In order to calculate the volume of the irregular shape, we apply the means of integration. The volume can be got by multiplying the area of each section by the corresponding height. The different thresholds in the SEM figure correspond to the area of pores on different section. If we make the threshold as small as possible, the volume of the irregular shape can be obtained. Symbol Yi stands for different thresholds. Suppose the area of the smaller loop at the gray level of threshold Yi is A_i, the volume of pore between these two thresholds is equal to:

$$\Delta V_i = (\frac{A_{i+1} + A_i}{2}) \times (\mathrm{Y}_{i+1} - \mathrm{Y}_i) \tag{1}$$

$$dV = \Delta V_i \tag{2}$$

According to the calculus theory, when the value of ΔY_i (i.e. $Y_{i+1} - Y_i$) is small enough, the values of A_{i+1} and A_i will be extremely close. This time, the pore volume calculated will extremely approximates the real volume. This method will be expected to have similar satisfactory result, if it is generalized into all the pores (as a whole). The total volume of the pores can be expressed by:

$$V_{3D} = \sum_{i=1}^{255} (\frac{A_{i+1} + A_i}{2}) \times (Y_{i+1} - Y_i) \tag{3}$$

Also, we can get the 3D porosity at arbitrary thresholds of gray level:

$$n_{3D} = \frac{\sum_{i=1}^{m} (\frac{A_i + A_{i+1}}{2}) \times (Y_{i+1} - Y_i)}{(Y_m - Y_0) \times S_A} \tag{4}$$

In this formula, m is the value of thresholds and S_A is the area of the pixels of selected regions, which is analyzed by Image-Pro Plus (IPP) software. The initial threshold Y_0 is equal to 0.

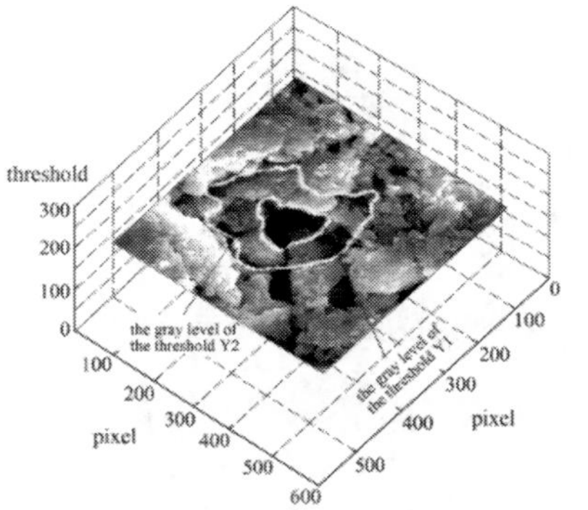

Fig.3. Calculate model of 3D porosity

Fig.4. Soil particles contact case

2.2 The rate of the contact area of clay particles

In principle of effective stress based on porous medium, Xu Xian-zhi [9] pointed out that the contact area of the soil particles is far less than that of the soil mass, which may apply to the soil with macropores or point-contacting particles. As for the soil with low porosity which has different kinds of contacting patterns of the soil particles, the contact area should not be ignored. Obviously, whether this conjecture is reasonable depends on the microstructure of the soil. In the microstructure figure of the soft clay shown as fig.4., contacting patterns of the soil particles is not merely point-contacting and not all porosities of the soil are large. So it's necessary to make further research to find out what the real contact area is, with deep insights into facts of contacting patterns of the soil particles.

Utilizing the statistical method presented in the paper, the 3D porosity of the micro figure can be obtained, comparing with the macro porosity which is got from the soil laboratory tests. When the values of these two porosities are the same, we define the corresponding pixel of the gray level analysis in the IPP figure as the contact area. And the ratio of the area of pixel to the total area of the pixel is defined as the rate of the average contact area (R_{CA}).

2.3 The relationship between the contact area and porosity

According to the method mentioned above, we carried out an experiment by scanning the soft clay from three cities (Hangzhou, Ningbo and Wenzhou) and got 225 SEM images analyzed, accomplished by configuring the moisture content of the soil to control the size of the porosity. The rate of the average contact area can be expressed by the statistical mean of 5 SEM images value from each test specimen with different porosity. We can get the relationship of the contact area (A_C), pore area (A_P) and total area (A) as following:

$$\frac{A_C}{A}+\frac{A_P}{A}=1 \tag{5}$$

Assuming $\frac{A_C}{A}=R_{CA}$, R_{CA} is the rate of the average contact area, based on average conception. The contact area includes the point contact, line contact and surface contact. Besides we define $\frac{A_P}{A}=R_{PA}$ as the rate of the average pore area.

Since the porosity and the rate of the average pore area are both representations of the soil pores, we can establish their relation as $R_{PA}=f(n)$. Then we can get the rate of the average contact area as following:

$$R_{CA}=1-f(n) \tag{6}$$

In practical situation, the soil is in a humid environment and the double electric layer on the surface of the clay particle includes the strong and weak bound water. The water film thickness of montmorillonite is 210Å (or Angstrom), and that of kaolin is 410 Å. The hydrophilic property of illitic soil is somewhere in between both of those, so its water film thickness ranges between 210 Å and 410 Å.

According to the definition of the length, we get 1Å $=10^{-10}$m$=10^{-4}$μm. On the principle of the microcosmic measurement, the length of 1 pixel is equal to 3.448μm when the amplification factor is 100 (Chen Cui-cui [10]). In the study, the amplification factor is set as 1000, so the length of 1 pixel is equal to 0.3448μm. In fig. 5., we can get the contact area of the clay considering the water film thickness:

$$A_C=A+S\times\frac{H\times10^{-4}}{0.3448} \tag{7}$$

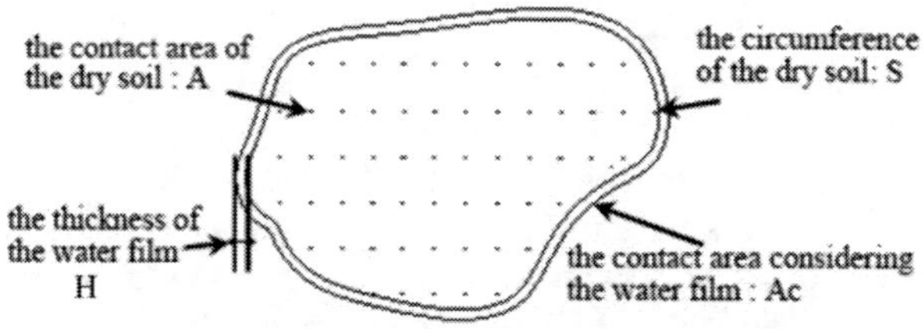

Fig. 5. The plan of particle contact area considering water film

Combining with equation (7), the relationship of the pore and the rate of the average pore area considering the water film thickness can be obtained. At the meantime, considering that the water film thickness varies with the properties of the soil, the water film thickness is set as 200 Å, 250 Å, 300 Å, 350 Å, 400 Å, and 450 Å respectively.

Statistical analysis is made to build the relationship of the porosity and the rate of average pore area, as shown in fig.6., fig.7. and fig.8..

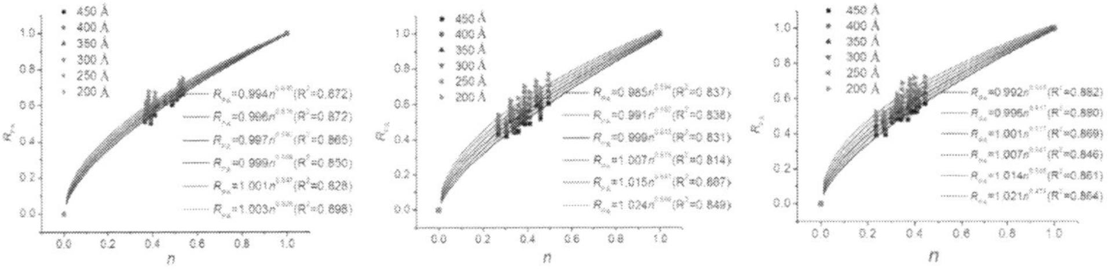

Fig. 6. ~ Fig. 8. The $n \sim R_{PA}$ curve of Hangzhou, Ningbo, Wenzhou soft clay in different combined water film thickness

We can get the conclusion from these figures that the rate of average pore area considering the water film thickness of the three are consistent in changing trends. With the porosity becoming larger, the rate of the average pore area increases. The results indicate that their relationship can be well fitted by a power function. The equation is as follows:

$$R_{CA} = 1 - R_{PA} = 1 - n^{\frac{2}{3}\alpha} \tag{8}$$

In the equation, R_{CA} is the rate of average contact area, and R_{PA} is the rate of the average pore area. α is an undetermined parameters, which is related to the composition and the district of the clay.

2.4 The static earth pressure considering the contact area of the particles by estimating water and earth pressures separately

Assuming the earth fill on the back of the retaining wall is porous soil, based on principle of effective stress, the effective stress is transferred through the contact points between particles of soil. The lateral pressure on the retaining wall only exists on the surface where the soil contacts with the wall. Ignoring the effect of pore water, the pores are not able to transfer effective stress. But if the pore is full with water and the retaining wall is waterproof, the pore water pressure can be transferred to the retaining wall through the pore structure. According to the effective stress theory considering the contact area of the particles, based on the results of microstructure researches in the paper, the total stress acting on the retaining wall on unit length is as follows, if the soil on the back of the retaining wall is uniform and above the underground water level:

$$E_0 = \frac{1}{2} K_0 \gamma H^2 (1 - n^{\frac{2}{3}\alpha}) \tag{9}$$

In the equation above, H is the height of the retaining wall, n is the porosity of the soil and $(1 - n^{\frac{2}{3}\alpha})$ is the rate of particle contact area according to the microstructure study in this paper.

In practice, the effect of buoyancy arising from groundwater on the back of the retaining wall should be taken into consideration when calculating the static earth pressure. Utilizing the method of estimating water and earth pressures separately and submerged unit weight of soil (γ') and considering the particle contact area and the pore area, we can get the soil pressure under water:

$$E_0 = \frac{1}{2}K_0 h_2^2(\gamma h_1 + \gamma' h_2)(1 - n^{\frac{2}{3}\alpha}) + \frac{1}{2}K_0 \gamma h_1^2(1 - n^{\frac{2}{3}\alpha}) + \frac{1}{2}\gamma_w h_2^2 n^{\frac{2}{3}\alpha} \tag{10}$$

When the porosity of soil goes to zero, which means the soil is elastic solid material, equation (9) and (10) are the conventional formula of soil pressure calculation.

3 Application example

Calculate the static soil pressure and the resultant force on the retaining wall in Fig.9., where $h_1 = 2\text{m}$, $h_2 = 4\text{m}$. The basic physical parameters of the soil are listed in Table 1.

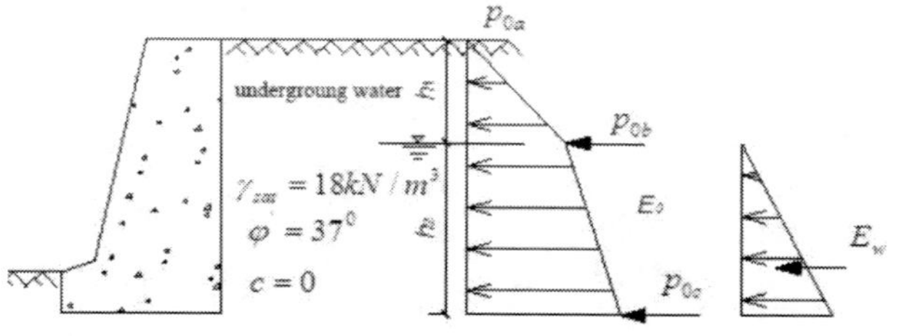

Fig. 9. Chart of earth pressure calculation

Table 1. Physical parameters of soft clay

g /(m·s^{-2})	γ /(kN·m^{-3})	γ_{sat} /(kN·m^{-3})	γ' /(kN·m^{-3})	c	φ /(°)
9.81	17.0	18.0	8.19	0	37

The coefficient of earth pressure at-rest K_0 is:

$$K_0 = 1 - \sin\phi' = 1 - \sin 37^0 = 0.4$$

(1) The conventional method of estimating water and earth pressures together

At point a, $p_{0a} = 0$

At point b, $p_{0b} = K_0\gamma h_1 = 0.4 \times 17 \times 2 = 13.6\text{kPa}$

At point c, $p_{0c} = K_0(\gamma h_1 + \gamma_{sat} h_2) = 0.4 \times (17 \times 2 + 18 \times 4) = 42.4\text{kPa}$

The resultant force by this method is:

$$E_0 = \frac{1}{2}(p_{oa} + p_{ob})h_1 + \frac{1}{2}(p_{ob} + p_{oc})h_2$$
$$= \frac{1}{2} \times (0 + 13.6) \times 2 + \frac{1}{2} \times (13.6 + 42.4) \times 4 = 125.6\text{kN/m}$$

(2) The conventional method of estimating water and earth pressures separately

At point c, $p'_{0c} = K_0(\gamma h_1 + \gamma' h_2) = 0.4 \times [17 \times 2 + (18 - 9.81) \times 4] = 26.7\text{kPa}$

The static water pressure on the wall:

$$E_w = \frac{1}{2}\gamma_w h_2^2 = \frac{1}{2} \times 9.81 \times 4^2 = 78.5\text{kN/m}$$

The resultant force is:

$$E_0 = \frac{1}{2}(p_{oa} + p_{ob})h_1 + \frac{1}{2}(p_{ob} + p'_{oc})h_2 + E_w$$
$$= \frac{1}{2}(0 + 13.6) \times 2 + \frac{1}{2}(13.6 + 26.7) \times 4 + 78.5 = 172.7\text{kN/m}$$

(3) Estimating water and earth pressures separately considering the contact area

As is shown above, the results of the former two methods are not associated with the porosity and the parameter of the contact area. So assuming the soil is the clay in Hangzhou the paper do tests on, we set $n = 0.4$, $\alpha = 0.8$.

$$E_0 = \frac{1}{2}(p_{oa} + p_{ob})h_1(1 - n^{\frac{2}{3}\alpha}) + \frac{1}{2}(p_{ob} + p'_{oc})h_2(1 - n^{\frac{2}{3}\alpha}) + E_w n^{\frac{2}{3}\alpha}$$
$$= \frac{1}{2}(0 + 13.6) \times 2 \times (1 - 0.4^{\frac{2}{3}\times 0.8}) + \frac{1}{2}(13.6 + 26.7) \times 4 \times (1 - 0.4^{\frac{2}{3}\times 0.8}) + 78.5 \times 0.4^{\frac{2}{3}\times 0.8}$$
$$= 84.57\text{kN/m}$$

Table 2. Results of calculation

Methods	**Resultant force** E_0 **/(kN/m)**
Estimating water and earth pressures together	125.60
Estimating water and earth pressures separately	172.70
Estimating water and earth pressures separately considering the contact area	84.57

The results above indicate that: the earth pressure reaches the maximum value which is 172.7kN/m utilizing the conventional method of estimating water and earth pressures separately, followed by the value utilizing the conventional method of estimating water and earth pressures together, which is 125.6kN/m, and the result of estimating water and earth pressures separately considering the contact area comes minimum, which is 84.57kN/m.

4 Conclusions

Comparing to the conventional method of forefathers' research, the method of estimating water and earth pressures separately considering the effect of the contact area of the particles on the lateral stress transfer, put forward in this paper, is consistent with the calculation of point stress but differs mainly in the stress calculation on contact surface. In the research, the influence of the material contact area on earth pressure transfer is studied, with more rigorous principles of stress transfer and more comprehensive theoretical formulas, which satisfies the stress transfer characteristics of granular materials more scientific. The formula has its preconditions and enforcing validity. Because water pressures act on both sides of the support in hydrostatic condition, they need not to be taken into account when calculating the anti-sliding and anti-overturning stability. Meanwhile, as for the conditions the excess pore water pressure or seepage water pressure can't be ignored, the influence that positive negative characteristic of the excess pore water pressure as well as that of the flow direction have on seepage pressure should be taken into consideration. For example, during the excavation of foundation pit, unloading effects lead to movement of retaining structures and deep-level strata heave, and in this case, the pore water pressure is negative.

Instead of using the strength reduction method to calculate the pore water pressure, the method put forward in this paper considers the real contact area of the pore from the structural features of clay itself, which corresponds closely to the actual conditions in a sense. But the soil is much more complicated actually, and the distribution of the pore scale may not always be uniform. On account that the influence coefficient of the contact area (α) has influencing factors within itself, how to calculate the lateral earth pressure more accurately is remained to be further researched and verified in numerous engineering practices. The rationalization mind of earth pressure on the support structure provides important guarantee for the engineering construction. Moreover, in addition to the safety and economy, more aspects such as multi engineering conditions need to be taken into account comprehensively during the design. Namely it saves investment, satisfies the demands of the engineering structure and really comes to achieve the double winning goal of social and economic efficiency.

This work is supported by the National Natural Science Foundation of China (Grant No. 41672264).

References

1. T. Li, B. Liu, Y. Li, et al, The void ratio of saturated red clay calculated from a micro-structure model, J. Journal of China University of Mining and Technology. **40(5)**, 720–725 (2011)
2. N.K. Tovey, D.H. Krinsley, Mapping of the orientation of fine-grained minerals in soils and sediments, J. Bulletin of Engineering Geology & the Environment. **46(1)**, 93–101 (1992)
3. Y. X. Wu. Quantitative approach on micro-structure of engineering clay, J. Bulletin of the Chinese academy of Geological Sciences. **23**, 143–151 (1991)
4. R. ENRIQUE, H.S. PAUL, Microstructure investigation in unsaturated soils: a review with special attention to contribution of mercury intrusion porosimetry and environmental scanning electron microscopy, J. Geotechnical and Geological Engineering. **26(6)**, 705–727 (2008)
5. J.C. BAKER, B. GRABOWSKA-OLSZEWSKA, P.J.R. UWINS, ESEM study of osmotic swelling of bentonite from Radzionkow, J. Applied Clay Science. **9(6)**, 465–469 (1995)
6. Y. J. Chen, Y.F Wen, Water and earth pressures on the supporting structure around a foundation pit, J. Chinese Journal of Geotechnical Engineering. **21(2)**, 139–143 (1999)
7. G. X. Li, Estimating the water and earth pressures on the supporting structure around a foundation pit separately and together, J. Chinese Journal of Geotechnical Engineering. **22(3)**, 348–352 (2000)
8. B. J. Wang, B. Shi, Y. Cai. et.al, 3D visualization and porosity computation of clay soil SEM image by GIS, J. Chinese Journal of Geotechnical Engineering. **29(1)**, 251–255 (2008)
9. X. Z. Xu, P. C. Li, C. L. Li, Principle of effective stress based on porous medium, J. Mechanics in Engineering. **23(4)**, 42–45 (2001)
10. C.C. Chen, W. Sun, W. L. Zhou, Quantitative image analysis on inner damage of cementitious material, J. Journal of the Chinese Ceramic Society. **4**, 574–580 (2010)

MATEC Web of Conferences **142**, 02011 (2018) https://doi.org/10.1051/matecconf/201814202011
ICMAE2017

Mineralogical Characteristics of a Micro-Fine and Low Grade Refractory Gold Ore

Yong Chen[1*], *Yongsheng* Song[1] , *Wenjuan* Li[1] and *Liulu* Cai[1]

[1]General Research Institute of Nonferrous Metals, National engineering laboratory of biological metallurgy, 100088 Beijing, P.R.China

Abstract. Detailed characterization of a micro-fine, low-grade refractory gold ore from a beneficiation plant in Yunnan Province, China, was investigated. Different analytical techniques, such as videlicet size analysis, scanning electron microscopy, and mineral analysis by a Mineral Liberation Analyser, were used for mineralogical characterization of the ore. The results show that the most valuable metal is gold. Grain size of pyrites and arsenian pyrites, which are main gold carrying minerals, are mostly 1 ~ 30 μ m. The gangue mineral is composed of quartz, mica, dolomite and feldspar, while the content of easy-to-float and clay gangue minerals is up to 33%.

1 Introduction

With the rapid development of the gold industry, refractory gold ores are becoming increasingly important resources for gold production [1,2]. Among these refractory gold ores, sulfur and micro-grained native gold ores represent a significant challenge, since they are not amenable to gold recovery by direct cyanidation. The gold contained in this type of ore is very finely disseminated and encapsulated in sulfides, which makes it inaccessible to the cyanide solution during direct cyanidation [3,4]. Autoclave pressure oxidation (AC POX) of the sulfide ore and subsequent cyanidation is a common technology used to recover the gold in such types of ore. Flotation is often used as a pre-treatment method to refractory gold ore [5].

As the gold-carrier minerals are usually finely disseminated (sometimes under 10 μm) and intimately associated with gangue minerals, a fine grind is necessary to get a high degree of liberation. At the same time, the fine grind usually causes the production of significant quantities of slimes, which comprise easy-to-float gangue minerals such as mica, dolomite, and calcite. These mineral phases are naturally hydrophobic and float readily, competing with the value minerals in the froth phase and resulting in lower concentrate grades.

Many researchers and companies have carried out studies on different kinds of ores using flotation. Marsden and House (2006) examined hydrogen peroxide in the selective flotation of auriferous pyrite with potassium amyl xanthate (PAX). They showed a high selectivity for auriferous pyrite when the pH exceeded 10 in the slurry, but with a minimal

* Corresponding author: cy040311@163.com

increase in recovery. Wang, Wei, Li, and Xie (2006) used the flowsheet of one rough-five cleaners-three scavengers with mixed collectors of butyl xanthate and butyl dithiophosphate on a refractory gold ore from Shuanglong Gold Mine of Shanxin, which contained arsenic and carbon with fine dissemination of gold particles. They achieved concentrate gold grade 36.04 g/t and gold recovery of 76.15%. For a low grade gold-bearing sulfide ore from Yunnan Province that contains 1.03 g/t Au, the valuable minerals were recovered by flotation using DPS reagent[6,7,8]. In micro-fine-grained gold ore containing arsenic and sulfur from the Hubei Province, it may be beneficial to involve measures of stage grinding-enhanced dispersion of slime-activated gold-carrier to acquire qualified Au concentrate[9].

Ore from the Zhenyuan gold mine in Yunnan represents a typical refractory gold ore. Under an optical microscope, the maximum size of the gold particles is only 8.5 μ m, while the average particle size is 2–5 μ m. The gold particles are coupled with gangue minerals containing a lot of secondary sericite of mud and clay minerals. The flotation index of the plant is low. In this study, tests were conducted using a Mineral Liberation Analyser (MLA) on the ores. The main objective was to find out the main gold-bearing minerals and dissemination size that influence flotation performance. Attention was paid to the occurrence of main metallic minerals and the mineral liberation degree of pyrite.

2 Materials and methods

A composite sample was collected from Zhenyuan gold mine at Ailaoshan district, Yunnan province. Approximately 200 kg of representative ore samples was crushed to –2 mm with a two-stage jaw crusher and one-stage roll crusher. The materials were then well mixed and divided into 1 kg samples for chemical and mineralogical studies. The sample was analyzed using atomic absorption spectrophotometry (AAS, GGX-9, Beijing, China).

3 Results and discussions

3.1 Chemical component of ore

The results of multi-element analysis of the sample are shown in Table 1. The result indicates that the most valuable metal in the ore is Au, while the harmful elements mainly are As and C. Table 2 shows the result of chemical phase analysis of Au in the ore. It demonstrates that sulfide and silicate are the main forms of inclusion gold mineral. The distribution rates of the two forms are 77.94% and 17.24% respectively. Single gold and gold trapped by the sulphide accounts for 80.26%, which is the maximum theoretical recovery of Au.

Table 1. Chemical analysis results of the feed ore.

Elements	Cu	Pb	Zn	S	Fe	Au (g/t)	Ag (g/t)
Content/%	0.032	0.14	0.025	0.87	3.87	1.16	<5
Elements	Ca	Mg	SiO_2	Al	C	As	Sb
Content/%	4.37	2.18	58.11	5.52	2.56	0.14	0.011

Table 2. Chemical phase analysis of Au in ore.

Phase	**Single/fissure**	**Carbonate**	**Metal oxide**	**Sulfide**	**Silicate**	**Total**
Mass fraction/%	0.018	0.02	0.009	0.913	0.20	1.16
Distribution rate/%	2.32	1.72	0.78	77.94	77.94	100.00

3.2 Mineral composition and content of ore

The mineralogical determination of the feed sample obtained with a Mineral Liberation Analyser (MLA) and back scattered electron imaging (SEM, JSM-6510, Japan, operating at 20 kV) is shown in Fig. 1. The analysis provided by the MLA is reported elsewhere in detail [10,11].

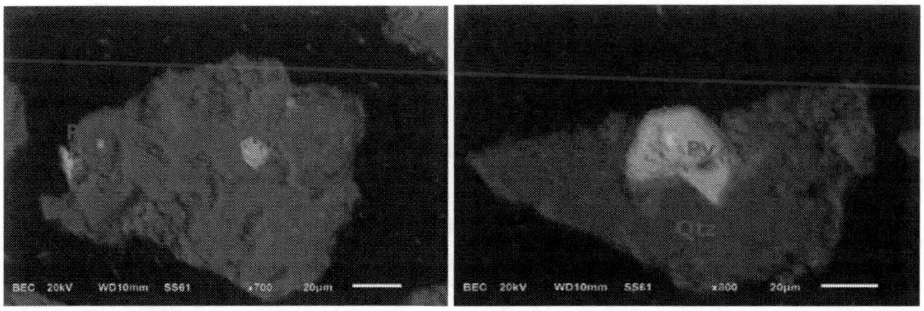

Fig. 1. Selected back scattered electron particle images of the sample illustrating the pyrite (Py) ore mineralogy and textures

From the chemical composition analysis and the MLA results (Fig. 2), it can be seen that the pyrite and quartz were the main sulfide and gangue minerals in the feed ore, respectively. However, the gold-carrier mineral (pyrite) was finely embedded. For the pyrite mineral phase, 61.08% of the pyrite grains were < 22 μm. In addition, the pyrite was closely associated with quartz, dolomite, calcite, and muscovite. As a result, the recovery of gold from the ore would be difficult due to the following features:
(1) The grades of gold in the ore were very low, only 1.16 g/t.
(2) The particle size was very fine, needing fine grinding, which is detrimental to the flotation process.
(3) The content of easy-to-float and clay gangue minerals was up to 33%, so the surface of a number of valuable minerals would be coated by slimes via electrostatic attraction; therefore, their surface hydrophobicity and subsequent flotation would be reduced.

3.3 Occurrence of main metallic minerals

Pyrite was the main gold-bearing minerals, some pyrite contained a small amount of chromium (Cr: 0 ~ 4.00% wt%) and arsenic (As: 0~30.97% wt%). The spectrum of pyrite and As-bearing pyrite were shown in Fig. 3. Pyrite accounted for 1.5% of the total mineral volume, and presented fine-grained mostly disseminated in the gangue, a small amount of vein distribution in the gangue fissure. Pyrite was mostly embedded in quartz, mica and feldspar, or partly wrapped in limonite, chromite, arsenopyrite and porphyrite. The maximum particle size of pyrite in the ore was 1mm while the minimum size was only 1~2 μ m, showing two grain size intervals.

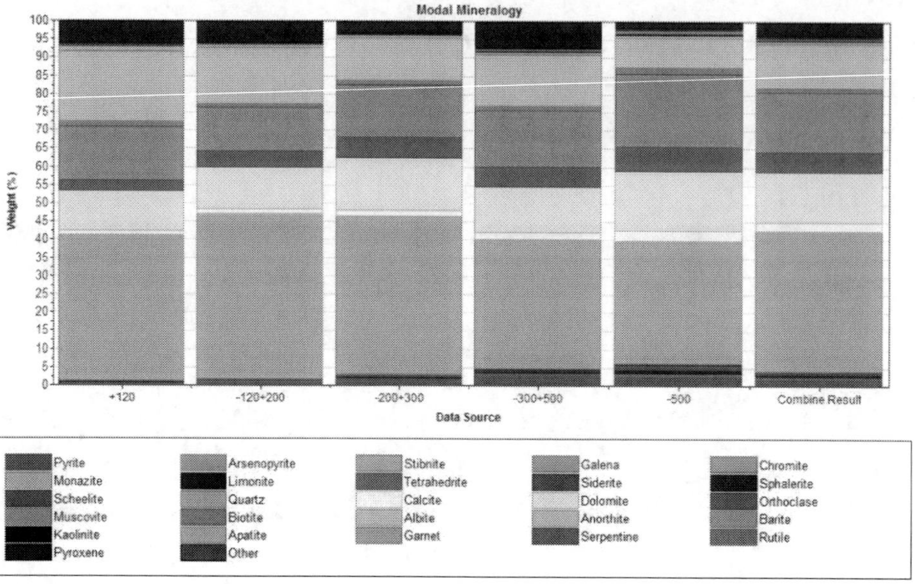

Fig. 2. The mineral composition and content result of the feed ore

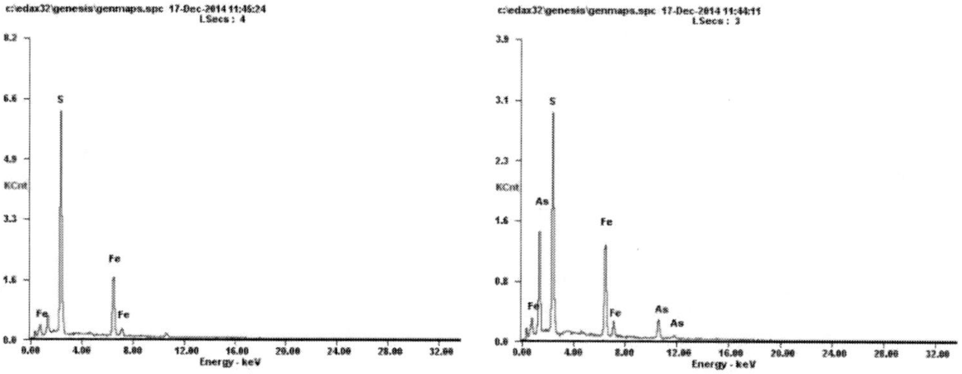

Energy spectrum diagram of pyrite

Energy spectrum diagram of arsenopyrite

Fig. 3. The energy spectrum diagram of pyrite and arsenopyrite

3.4 The main mineral dissociation degree and the relationship

The relationship of different minerals at a grinding fineness of 70% -200mesh was shown in table 3. Pyrite free surface was covered for 75.36%, 7.42% contacted with quartz, 5.97% contacted with muscovite, 3.05% contacted with dolomite, 3.61% contacted with albite. These minerals were main gangue minerals associated with pyrite.

Table 3 Interlocking relation between main mineral and other minerals

Mineral	pyrite	arsenopyrite	stibnite	galena	tetrahedrite	monazite	chromite	limonite	siderite
pyrite	——	2.57	0.20	0.00	17.21	0.00	0.86	0.71	0.03
arsenopyrite	0.42	——	0.00	0.00	0.00	0.02	0.13	0.58	0.01
stibnite	0.00	0.00	——	0.00	0.00	0.00	0.00	0.00	0.00
galena	0.00	0.00	0.00	——	0.00	0.00	0.00	0.00	0.00
chromite	0.10	0.13	0.00	0.00	0.00	0.00	——	1.93	0.42
monazite	0.00	0.00	0.00	0.00	0.00	——	0.00	0.00	0.00
limonite	0.30	3.29	0.00	0.00	0.00	0.00	9.52	——	0.69
tetrahedrite	0.13	0.00	0.00	0.00	——	0.00	0.00	0.00	0.08

siderite	0.03	0.03	0.00	0.00	9.61	0.00	3.69	0.99	——
sphalerite	0.00	0.00	0.00	0.00	5.68	0.00	0.00	0.00	0.00
rutile	0.02	0.00	0.00	0.00	0.00	0.00	0.00	0.04	0.00
quartz	7.42	1.76	0.04	0.51	1.19	0.92	4.14	5.63	6.95
calcite	0.40	0.00	0.00	0.00	0.00	0.00	0.19	0.06	0.05
dolomite	3.05	3.60	0.00	0.00	1.83	0.00	3.70	1.78	4.48
orthoclase	1.21	0.14	0.00	0.00	0.95	2.19	0.60	1.67	0.97
muscovite	5.97	4.02	0.00	0.00	8.64	23.53	5.02	7.38	9.96
biotite	0.26	0.16	0.00	0.00	0.00	0.00	1.29	2.27	1.56
albite	3.61	1.12	3.57	0.00	4.04	1.69	2.29	2.72	3.24
anorthite	0.22	0.10	0.00	0.00	0.66	0.00	0.09	0.41	0.38
barite	0.02	0.00	0.00	0.00	0.00	0.00	0.00	0.00	0.00
kaolinite	0.01	0.00	0.00	0.00	0.00	0.00	0.05	0.87	0.13
apatite	0.09	0.00	0.00	0.00	0.00	0.00	0.07	0.02	0.03
garnet	0.13	0.00	0.00	0.00	0.00	4.49	0.09	0.02	0.03
serpentine	0.02	0.00	0.00	0.00	0.00	0.00	0.36	0.07	0.16
pyroxene	1.23	0.52	0.00	0.00	0.51	0.00	0.09	2.18	0.39
others	0.00	0.00	0.00	0.00	0.00	0.00	0.00	0.00	0.13
free surface	75.36	82.57	96.18	99.49	49.67	67.10	67.83	70.66	70.29
Total	100.00	100.00	100.00	100.00	100.00	100.00	100.00	100.00	100.00

4 Conclusions

Mineralogical analysis of the feed using MLA indicated that pyrite was the main gold host mineral in the ore. Quartz, muscovite, and dolomite were identified as the principal gangue minerals. The total content of easy-to-float and clay gangue minerals was up to 33%. Pyrite was the main gold carrying mineral but finely embedded, with almost 61.08% of the pyrite particles being < 22 μm. the recovery of gold from the ore would be difficult.

The authors are grateful for the financial support from National Natural Science Foundation of China (No.51504031) and China Gold group for this work. We would like to express our sincere appreciation to the anonymous reviewers for generous advice and help for language modification

References

1. O' Connor C.T., Dunne R.C., The flotation of gold bearing ores-a review. Min. Eng., **7**, 839 (1994)
2. Agorhom, E. A., Skinner, W., & Zanin, M. Upgrading of low-grade gold ore samples for improved particle characterisation using Micro-CT and SEM/EDX. Advanced Powder Technology, **23**, 498 (2012)
3. Allan, G.C., & Woodcock, J.T. A review of the flotation of native gold and electrum. Minerals Engineering, **14**, 931 (2001)
4. Bulatovic, S.M. Flotation behaviour of gold during processing of porphyry copper–gold ores and refractory gold-bearing sulphides. Minerals Engineering, **10**, 895.(1997)
5. Monte, M.B.M., Lins, F.F., Oliveira, J.F., Selective flotation of gold from pyrite under oxidizing conditions. Int. J. Miner. Process. **51**, 255(1997)
6. Bradshaw, D.J., Harris, P.J., & O' Connor, C.T. Synergistic interactions between reagents in sulphide flotation. J. S. Afr. Inst. Min. Metall., **98**, 187 (1998)
7. Wang, X.F., Wei Y.R., Li J.B., Xie J.H., Flotation of refractory gold ore containing arsenic and carbon with fine gold dissemination from shuanglong gold mine in Shanxi of china. Nonferrous Metals. **6**, 63(2006)

8. Jue, S.J., Wang, H.F., & Zhao, F. Beneficiation experimental investigation on a low grade gold-bearing sulfide ore. Yunan Metallurgy, **1**, 18 (2012)
9. Lin F., Li X.Y., Xiao Y.Y., Li Y., Study on Flotation Test of Microfine-grained Gold Ore Containing Arsenic and Sulfur. Resources Environment& Engineering, **4**, 192 (2009)
10. Gu, Y. Automated scanning electron microscope based mineral liberation analysis. Journal of Minerals and Materials Characterization and Engineering, **5**, 2(2003)
11. Fandrich, R., Gu, Y., Burrows, D., & Moeller, K. Modern SEM-based mineral liberation analysis. International Journal of Mineral Processing, **84**, 310(2007)

A Measuring Method of Gravel's Coefficient of Restitution and Discussion

Yanhui Chen[1,*], *Xinjie* Zhang[1] and *Bin* Wang[1]

[1.] School of Mechanical Engineering, Guangxi University of Science and Technology, 268 East Central Avenue Liuzhou, China

Abstract. In this paper, a simple method for measuring the recovery coefficient of gravel was proposed by the definition of the coefficient of restitution and the principle of kinematics. The recovery coefficient between gravel and slate was obtained by the experiment; meanwhile the reason of error was analyzed and discussed. Finally, the experimental results were compared with the simulation data of the EDEM software. It was concluded that the data were more appropriate when the gravel particles were smaller.

1 Introduction

The material properties were required to have the material recovery coefficient, which was in many of the simulation software. The coefficient of restitution of the material used in particular was a certain gap found in the book. Meanwhile, the cost of the equipment measured by traditional method was high (measuring the trajectory of the particulate material by the high speed camera). Therefore, this paper provided a simple method of measurement. At the same time, this paper also discussed and analyzed to find the error of this method and the measurement accuracy of the recovery coefficient.

2 The coefficient of restitution

The coefficient of restitution refers to the recovery of a physical quantity when two objects collided [1]. Newton first introduced the definition of collision recovery coefficient in the classical mechanics. He defined the ratio of the velocity at which the two objects were about to be separated after the collision and the velocity before the collision would be in order to define the coefficient of restitution e.

$$e = \frac{v_1 - v_2}{v_{10} - v_{20}} \tag{1}$$

In the Eq. (1), $v_1 - v_2$ as the velocity at which the two objects were about to be separated after the collision; $v_{10} - v_{20}$ as the velocity before the collision.

* Corresponding author: gxut_jx@163.com

For the study of the coefficient of restitution, the influence of the collision angle of the recovery coefficient, the experimental method and the significance of the restitution factor were studied in literatures [2-4]. In the literature [5], a new kind of device for measuring the coefficient of restitution was proposed. In this paper, a simple recovery coefficient measuring device was designed by studying the coefficient of restitution and using the principle of kinematics and the definition of the coefficient of restitution in the classical mechanics.

3 Introduction of experimental device

Experimental device was shown in Fig.1.

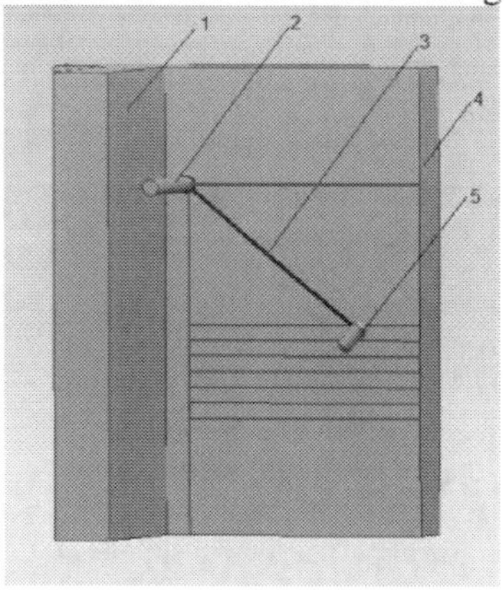

Fig. 1. The experimental device

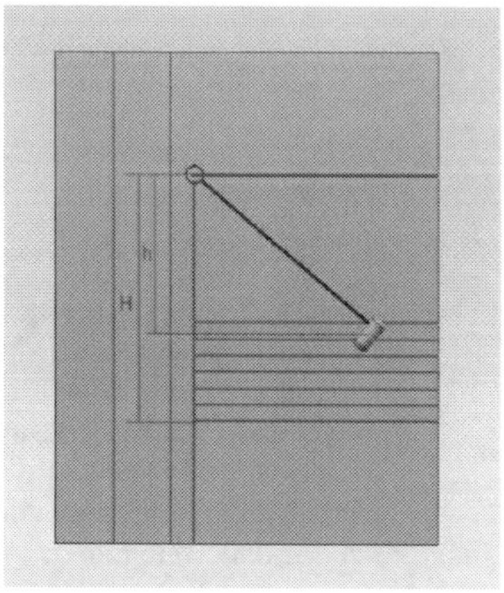

Fig. 2. The experimental principle

In the Fig.1:
1 — the collision plate (can change materials);
2 — the support shaft;
3 — the string (the lower part of it tied to the sample, the upper part of it tight support shaft, the material requirements must be no elasticity);
4 — the scale plate (the lower part of it was the length measuring unit, the maximum range was 35 cm, the horizontal line was the scale line, and the red line was 0 scale line).
5 — the collision sample;

This experimental device measured the recovery coefficient of the material in the form of a positive collision, so it was necessary to ensure that the direction of movement of the sample was perpendicular to the height of the collision plate when the sample collided with the collision plate.

3.1 Experimental description

First, the sample was tightened with the string to measure the distance between the support shaft and the center of gravity of the sample at this time. Held the sample and straightened the string and coincided with the 0 scale line. Then released the sample, so that the sample was freed to fall. At this point, the drop height was the distance from the support shaft to the center of gravity of the sample. The sample bounced to a certain height after colliding with the collision plate. At this time there may be various reasons leading to that the rotation of the sample can't be bounced or bounced after the collision to the scale plate, this can't be successful. Only the trajectory of sample was perpendicular to the plane of collision plate and there was no obvious rotation when the sample bounced, then it was considered successful. Repeated this process several times and recorded this process with the video camera. Later, the video processing software be used after slow release, 1s can be divided into 25 frames for processing, read the successful bounce height.

3.2 The principle of experiment

In the experiment, without considering the mass of the string, the friction loss between the string and the support shaft and between the string and the sample, only gravity work was down in the whole process. Therefore, the velocity of the sample at the collision met the energy conservation.

$$\frac{1}{2}mv_{10}^{\ 2} = mgH \tag{2}$$

In the Eq. (2), m is the sample weight, H is the distance from the center of gravity of the sample to the support shaft in the vertical direction, g is the acceleration of gravity and v_{10} is the velocity of the sample before the collision. After the collision, the velocity after the collision can be calculated according to the energy conservation.

$$\frac{1}{2}mv_1^{\ 2} = mg(H-h) \tag{3}$$

In the Eq. (3), h is the vertical distance from the lowest point after the bounce, v_1 is the velocity of the sample after the collision.

Because the velocity of the plate has been 0 before and after the collision, so simultaneous Eq. (1) (2) (3) can be obtained Eq. (4).

$$e = \sqrt{\frac{H-h}{H}} \tag{4}$$

3.2.1 Analysis of the device error

The possible error of the device in measuring the coefficient of restitution of the material was analyzed as follows:

(1) The friction between the string and the support shaft. This friction would consume mechanical energy and transformed it into internal energy, but there was the rolling friction, so bold speculation was less effective and can be ignored.

(2) The friction loss between the string and the sample can be ignored basically, as long as the binding was firm and the displacement was not generated.

(3) A slight rotation was produced even in the case of success, resulting in a lower height of the bounce.

3.3 Experimental verification

3.3.1 The coefficient of restitution of collision between gravel and slate was studied

In order to verify the feasibility of the experiment device, the coefficient of restitution between the gravel was taken as an example to carry out the experiment. At the same time, the parameters be measured which were simulated by EDEM software and compared with the results of the cylinder experiment.

3.3.2 The coefficient of restitution of the gravel was measured

In this experiment, three pieces of ordinary gravel of large, middle and small three types were selected as A1, A2, A3, B1, B2, B3, C1, C2, C3 and an ordinary slate as the collision material, which was measured with the experiment device. Among them, the initial height of large, middle and small three kinds of gravel were 32 cm, 31.5 cm and 31 cm respectively. The size of the gravel was shown in Fig.3.

Fig. 3. Gravel size

The measured data were shown in Table 1.

Table 1. Data measured by the device

Sample	Group	Initial height [cm]	Bounce height [cm]	Coefficient of restitution
	1		26.7	0.407
	2		27.3	0.383
A1	3		26.9	0.399
	4		26.1	0.429
	5		26.3	0.422
	1		27.4	0.379
	2		26.5	0.415
A2	3	32	26.3	0.422
	4		26.1	0.429
	5		26.0	0.433
	1		26.5	0.415
	2		25.1	0.464
A3	3		27.1	0.391
	4		25.2	0.461
	5		25.5	0.451
	1		26.1	0.414
	2		25.7	0.429
B1	3		25.1	0.451
	4		26.8	0.386
	5		25.3	0.444
	1		25.7	0.429
	2	31.5	24.3	0.478
B2	3		24.8	0.461
	4		26.0	0.418
	5		24.1	0.485
	1		24.3	0.478
	2		25.1	0.451

B3	3		26.8	0.386
	4		26.1	0.414
	5		24.7	0.465
	1		23.0	0.508
	2		21.1	0.565
C1	3		22.5	0.523
	4		24.2	0.468
	5		23.3	0.498
	1		25.1	0.436
	2		23.9	0.478
C2	3	31	25.3	0.429
	4		23.1	0.505
	5		23.3	0.498
	1		23.4	0.495
	2		24.1	0.472
C3	3		22.8	0.514
	4		21.8	0.545
	5		26.5	0.559

Through the experimental analysis of the above data, we can draw the following conclusions: 1. The experimental data had obvious gap for the different gravel, even with the same gravel also had a big difference. As for the reasons, the author thought that on one hand it was the individual difference in the shape and the mass of the different gravel; On the other hand, the same gravel was from different contact point of the collision plate during the collision [6]; 2. For the larger particles of gravel, the values of the recovery coefficient were removed by the obvious large or small values, and the rest values were basically between 0.38 and 0.43, and the average of this part was 0.41 by calculation; For intermediate particles of gravel, the recovery coefficient was basically concentrated in 0.41 to 0.48, and the average of this part was 0.447 by calculation; For smaller particles of gravel, the values of the recovery coefficient were removed by the obvious large or small values, and the rest values were basically between 0.46 and 0.55, the average of this part was 0.50 by calculation; 3. Different results had been obtained for different sizes of gravel, which fully demonstrated the size of the gravel in this experiment device had a significant impact on the results of the experiment.

3.3.3 The repose angle was measured by cylinder test

The bottomless cylinder experiment was one of the many experiment devices for measuring the repose angle. The bottomless cylinder be placed in a horizontal plane with the gravel filled cylinder, and then lifted cylinder slowly and vertically, the gravel was free to fall under the action of the gravity, then a cone formed by natural flow. In this experiment, the cylinder with the aspect ratio of 1.5:1 was used to fill the gravel with randomly, and it was lifted with uniform speed by the hoist. The gravel formed the rest angle on the ground covered by PVC plate, then took a picture perpendicular to the ground and measured the rest angle in the picture.

Measurement of experimental data was shown in Table 2.

Table 2. The rest angle of cylinder test measured

Sample Classification	Rest Angle [°]	Average [°]
1	27.6	
2	26.4	
3	26.6	26.5
4	28.1	
5	26.2	

6	23.9	
7	27.8	
8	25.6	
9	26.8	
10	26.1	

3.4 EDEM simulation

Li et al. [7] showed that it was proved that the parameter calibration in EDEM software can predict the motion of particles correctly by the rest angle experiment under the different devices with the ball made of different materials and compared with the simulation results of EDEM software. The size of the rest angle was calculated by simulating the three coefficient of restitution measured above.

A bottomless cylinder model was established with the radius of the bottom circle being 100 mm, and the height being 500 mm (the height of the cylinder was raised so as to facilitate the generation of particles), the material was set as PVC. The ground area was set to 1000 mm * 1000 mm, and the material was PVC. The shape of gravel was irregular, so two kinds of irregular particles were set up. Its shape was shown in Fig.4, each of the two models consists of four different circular particles, in which the maximum diameter of particle 1 was 18.5 mm, and the maximum diameter of particle 2 was 20.5 mm. When the particles were generated, the gradation measured according to the above tests were divided into four levels of size, which were 1 times, 1.5 times, 2 times and 2.5 times the original particle size respectively. The weight measured under each gradation of experimental gravel was used as the standard for the generation of the individual particle levels, which were 6.36 kg, 2.86 kg, 2 kg and 1.23 kg respectively. The lifting speed of the cylinder was measured actually by 0.06 m/s, the cylinder moved in a straight line in the positive direction of the Z axis after 0.6 s due to the early stage of particles generation to a steady state needed 0.5 s, until the end of the simulation. The values of the rest angle were obtained by the screenshot measurement. The simulation parameters were shown in Table 3.

Table 3. Simulation parameters

Material	Parameters					
	Density [kg·m-3]	Shear modulus [GPa]	Poisson's ratio	Sliding friction coefficient	Rolling friction coefficient	Coefficient of restitution
Gravel	2090	11.1	0.29			
PVC	1400	1.45	0.35			
Gravel and the ground(PVC)				0.47	0.002	0.45
Gravel and gravel				0.84	0.004	A: 0.41 B: 0.45 C: 0.50
Gravel and cylinder wall (PVC)				0.47	0.002	0.45

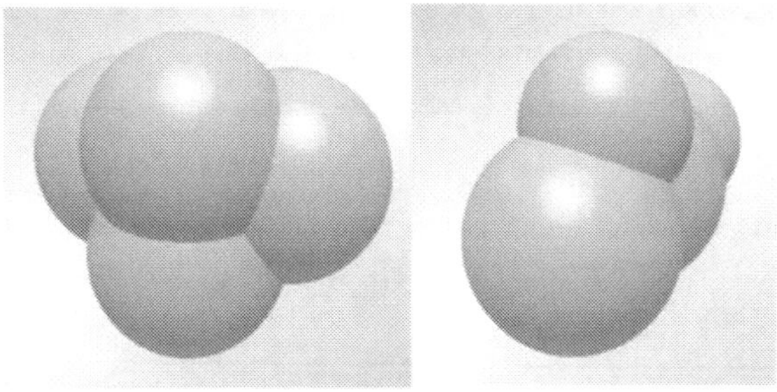

Fig. 4. Particle model

Fig. 5. Image contrast

3.4.1 The contrast between simulation results and experimental results

The results obtained by repeated simulation were shown in Table 4. Fig.5 was a comparison of the experiment and the simulation.

Table 4. The results of simulation data

Group	Rest Angle [°]	Group	Rest Angle [°]	Group	Rest Angle [°]
A1	31.5	B1	26.5	C1	25.6
A2	30.2	B2	28.2	C2	26.0
A3	28.8	B3	26.2	C3	27.2
A4	32.2	B4	30.4	C4	28.3

In the Table 4, The recovery coefficient of the large particles in Group A was bigger than the simulation results, and the error was larger with the experimental results. The experimental results of Group B and Group C were smaller error compared with Group A. It can be seen from Fig.5, the Group B and C were more similar to the experiment, but the simulation data of Group C were basically consistent with the experimental data, and the part values of the Group B were too large, therefore the recovery coefficient of the Group C were more appropriate.

4 Conclusions

Through the comparison, we found that the recovery coefficient of the small particle was the most suitable as the simulation parameters. The large particle measured values smaller, the author thought that the larger the gravel was, the greater its mass was, and the greater the tension on the string, resulting in the greater friction loss between string and support shaft, and the greater friction loss between string and the large particle of gravel, resulting in a smaller measurement result eventually. And, the smaller particles had little effect because of lighter.

The method proposed in this paper can accurately and conveniently measure the coefficient of restitution of collision between the materials. The comparison experiment shown that when the smaller particles were chosen, the experiment results were closest to the theoretical data and the simulation results were more accurate.

This work is supported by Guangxi Science and Technology Development Program Funded Projects: The construction of the comprehensive test platform about earthmoving machinery operation process (No. GuiKe 1598021-2).

References

1. C. L. Wang. The discussion of the coefficient of restitution. University Physics, **16**, 3(1997)
2. Y. Wang. The discussion of the one-dimensional head-on collision coefficient of restitution. The Physical and Engineering, **24**, 4(2014)
3. S. M. He, Y. Wu, and X. P. Li. Rolling impact study coefficient of restitution. Rock and Soil Mechanics, **30**, 5(2009)
4. C. H. Xu. The meaning and scope of coefficient of restitution. Physical Bulletin, **4**, 2(1998)
5. M. J. Yang, L. Yang, and Q. D. Li. Simple granular material coefficient of restitution measurement method and its application. Agricultural Mechanization Research, **10**, 3(2009)
6. Z. Y. Qin and Q. S. Lu. Based on the coefficient of restitution of collision simulation. J. Dynamics & Control, **4**, 5(2006)
7. Y. J. Li and Y. Xu. Colin Thornton. A comparison of discrete element simulations and experiments for 'sandpiles' composed of spherical particles. Powder Technol., **160**, 10(2005)

Failure Analysis and Safety Protection of a Certain Type of Wire Ropes under High-Speed Impact Loads

Zhiqian Ren[1]*, *Zhiqiang* Lu[2], *Qiong* Yu[2] and *Yun* Jiang[2]

[1]Science and Technology on Integrated Logistics Support Laboratory, National University of Defense Technology, Changsha, China
[2]Hunan Sushiguangbo Testing Technology Co.,Ltd, Changsha, China

Abstract. As the key bearing components in engineering projects, wire ropes often suffer from high-speed impact loads in service. In the majority of applications, wire ropes are subjected to the loads with the following characteristics: alternating impact loads, short-time overload, and small movements among adjacent spiral strands. Typical failure modes of wire ropes include fracture, wear, corrosion, geometric failure and thermal failure. In this work, the corresponding failure mechanisms were analyzed. For a certain type of wire ropes, the fracture morphology, surface quality, microstructure and internal defects of steel wires were analyzed, and failure causes were discussed. Accordingly, the precautions in relation to the safe use of wire ropes were proposed.

1 Introduction

Wire ropes are widely used in many industrial fields because of their large bearing capacity, good flexibility and smooth motions without noise. However, they tend to degrade due to a range of factors including environmental corrosion, uncertain alternating loads, mechanical impacts, wear and high-speed impact stress, resulting in wear, corrosion and wire fracture, hence decreased strength. With the reduction of strength, wire ropes may suddenly break, inducing security incidents such as production pauses and even casualties. Therefore, conducting the failure analysis of wire rope is of vital significance on both safety and economy in production.

2 Characteristics of wire rope loads

2.1 Alternating impacts

Each working cycle of wire ropes includes linear lift, movement through the pulley, coiling on the winding drum and unload drop. Therefore, the force on each steel wire is essentially

*Corresponding author: rzq_rtl@nudt.edu.cn

alternating stress, and the wire ropes are subjected to fatigue loads and alternating high-speed impact loads [1].

2.2 Short-time overload

Wire ropes elongate slightly with the increase of loading, and it may fracture when the load exceeds the elastic limit. Static loads of wire ropes are generally controlled within around 10%~20% amount of the minimum breaking strength, referred to as safe working loads. However, wire ropes are in a state of motions during use. In addition to the cargo weight, rope self-weight and other static loads, wire ropes also bear the dynamic loads caused by impacts and the acceleration of movements, the additional load caused by bending, and the resistance load caused by friction, etc. The actual safety factor would decrease when other loads apart from static loads increase, and wire ropes may be subjected to short-time overload in consequence [2].

2.3 Small movements among adjacent spiral strands

The structure of a typical wire rope is shown in Fig. 1, where the spiral strands are mutually twisted. The steel wires in the same spiral strand deform synchronously, but the deformations of adjacent spiral strands are slightly different, inducing relative sliding among strands and wires, which means that the steel wires are under fretting wear environment in actual service [3].

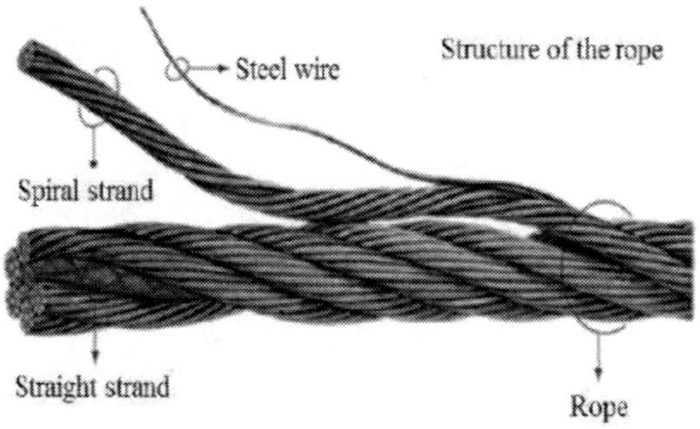

Fig. 1. Schematic diagram of the structure of a typical wire rope.

3 Typical failure modes and failure mechanisms of wire ropes

Some scrapped wire ropes were detached and analyzed, and the failure modes of wire rope were classified into three categories according to the external form, i.e., fracture, excessive deformation and surface damage. These failure modes and the corresponding failure mechanisms are summarized and discussed as follows.

3.1 Fracture

Fracture is the most common failure mode of wire ropes. Fracture failure can be divided into three kinds according to the special structure of wire ropes, i.e. fracture of steel wires, fracture of spiral strands and fracture of wire ropes. Fracture of wire ropes and spiral strands result from steel wire fracture, and therefore analyzing the failure mechanisms of steel wires fracture is the key to reveal the fracture of wire ropes.

Failure modes of steel wire fracture were widely studied in previous studies, and were classified into three kinds [4], i.e., (1) fracture from a single applied load, which is induced when a single applied load exceeds the bearing capacity of steel wires; (2) fatigue fracture resulting from alternating stress; (3) fracture from environmental mediums, which is low stress brittle fracture caused by environmental medium and applied stress.

Typical fracture morphologies are summarized from a great deal of fractured steel wires as shown in Fig. 2. Fracture from a single applied load shows tensile or shear breakage characteristics. A steel wire would fracture when the axial load exceeds its breaking strength, and the wire near the fractured region would elongate with the reduction of its diameter, resulting in the typical cup-cone fracture morphology, as shown in Fig. 2(a). Shear fracture is induced by the load perpendicular to the axial direction, and the angle between the fracture surface and the axis is 45°, as shown in Fig. 2(b). Fatigue is also commonly found to be the main reason for fractured steel wires. In traction and hoisting wire ropes, the fatigue fracture of steel wires is mainly caused by alternating bending under stress through the pulley and winding drum in consequence of the fatigue of metals. The fatigue fracture usually occurs among the outer steel wires in the worst bending side, and the fracture morphology is relatively flush, as shown in Fig. 2(c).

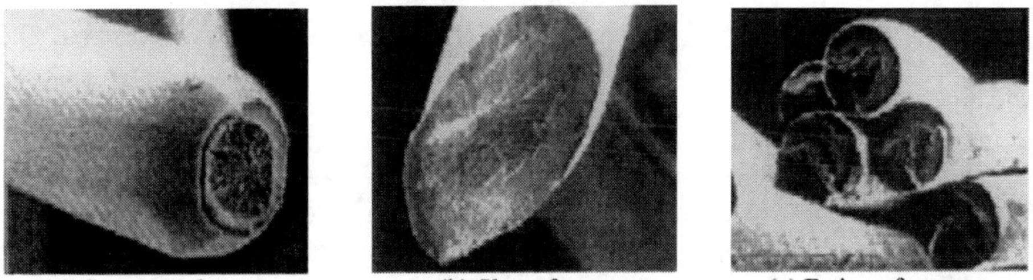

(a) Cup-cone fracture (b) Shear fracture (c) Fatigue fracture

Fig. 2. Typical fracture morphologies of fractured steel wires.

3.2 Wear

There are micro relative movements on the contact surface between the surface of a steel wire and pulley, winding drum or the adjacent steel wires, inducing the separation of surface material particles from the substrate due to the integrative effect of mechanical, physical and chemical processes. This would further lead to the change of shapes, dimensions and weight of wire ropes, i.e. wear of wire ropes. The phenomena of dimension reduction, surface state changes and even loss of functions are all classified as wear failure. Wear failure is a gradual process, which is not as sudden as fracture failure accidents. However, wear usually results in the decrease of bearing areas and the increase of fatigue sensitivity, which may further cause fracture failure, especially in corrosive mediums, as wear can accelerate corrosion. The wear failure can be classified into two kinds [5], i.e., mechanical wear and plastic wear. The main difference between the two forms is that the latter shows the material deformation or displacement without material loss phenomenon.

3.2.1 Mechanical wear

The reason for the mechanical wear of external steel wires lies in the fact that the steel wire constantly contacts with rope race, lifting hook, ground, etc. The mechanical wear is especially obvious on the contact areas between the wire ropes and pulley under acceleration and deceleration operations, leading to the decrease of wire diameters, loading areas and capacity. In addition, lack of lubrication, existence of sand and gravel grains between spiral strands can also aggravate the mechanical wear.

3.2.2 Plastic wear

Surface wear of wire ropes caused by vibration, impact and internal squeezing is called plastic wear, i.e. localized wear. Strikes from other objects on the wire rope coiled on a winding drum and mutual winding of wire ropes would lead to plastic wear. With wire ropes repeatedly curved, the interaction forces and slips between steel wires would occur, and then the contact stress between spiral strands increases, leading to localized indentation and plastic wear of wire ropes [6]. Although the cross-sectional area decreases slightly, surface materials in the localized squeezing place would harden, which tends to induce cracks or even fracture.

3.3 Corrosion

As wire ropes mainly serve in relatively severe working environments such as outside and marine environments, corrosion occurs due to the reactions with environmental mediums. Fracture, wear and corrosion are the main damage types for wire ropes. However, unlike fracture, both of wear and corrosion are gradual processes, and they usually interact with each other, inducing early failure of wire ropes. Meanwhile, corrosion can provide conditions for fracture, and accelerates or even directly causes fracture.

Wire rope corrosion can be classified into two categories [7]. The first kind is general corrosion due to contact with air, leading to chemical corrosion and electrochemical corrosion. Chemical corrosion occurs when wire rope is exposed in acidic environment, and electrochemical corrosion occurs in humid air environment. The second kind is localized corrosion such as pitting corrosion, by which holes or cavities are produced in the place where the coating breaks down and comes off. The corrosion pit gradually deepens on the wire rope surface, inducing stress concentration, fatigue crack and decrease of strength and flexibility. The stress concentration will accelerate the extension of cracks, and finally result in mechanical rupture.

3.4 Other failure modes

3.4.1 Geometric failure

Geometric failure mode occasionally occurs to wire ropes because of their special structures, as shown in Fig. 3.

The phenomena of loose strand and knotting are usually found in hoisting wire ropes, as shown in Fig. 3(a) and Fig. 3(b). Rotation against the twist direction and twist around its axis are the most common reasons for the failure [8].

Steel wire extrusion is also a common geometric failure mode, as shown in Fig. 3(c). It is caused by insufficient interspace between steel wires in a spiral strand. The steel wire cannot move with the adjacent wires when the interspace is insufficient, leading to overload or yielding. The interspace of outside spiral strands will be insufficient if the diameter of the rope core is too small, resulting in the damage of spiral strands [9].

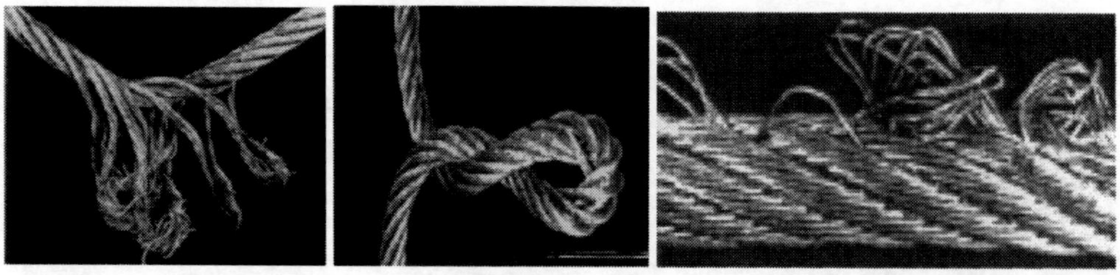

(a) loose strand (b) knotting (c) steel wire extrusion

Fig. 3. Typical geometric failure of wire ropes.

3.4.2 Thermal failure

Thermal failure mainly results from the formation of martensite and heat corrosion. Martensite is a brittle metal structure, a very thin layer of which can be usually found on the convex part of the wire. Electric arc caused by electric welding will induce the formation of martensite on the wire rope surface. This martensite layer tends to fracture when the wire rope bends, because it would lead to stress concentration, strength reduction and the formation of fatigue crack [10]. The fatigue crack extends rapidly, inducing abrupt fracture of wire ropes in service.

As it is a good conductor of heat, localized temperature of wire ropes will increase when it works under the environment of high temperatures. Recrystallization would occur to the microstructure of cold drawn wire ropes if the internal temperature exceeds 300℃, and then the tensile strength of steel wires would drop to two-thirds of the value at room temperature [11]. Wire ropes will be heated up fast if the heat absorption exceeds thermal dissipation, leading to the heat corrosion.

4 Failure analysis cases of wire ropes

4.1 Fracture morphology of steel wires

Scanning electron microscope (SEM) morphologies of typical fractured steel wires are shown in Fig. 4. The fracture mechanism of the wire rope is determined to be fatigue fracture judging from the morphologies of the fracture surfaces. A small fraction of fractured steel wires in the service of a wire rope would not cause the abandonment of the whole wire rope. If some fractured steel wires that were not cut out in time, and the rupture region would consequently be rubbed to a flat area, as shown in Fig. 4(a). Torn areas can still be clearly observed in some fatigue fracture surfaces, as shown in Fig. 4(b) and Fig. 4(c). Through analyzing the fracture morphologies of the steel wires, it is revealed that most fatigue fracture is caused by high nominal stress, and stress concentration exists in crack source on a small part of the fracture surface, as shown in Fig. 4(c).

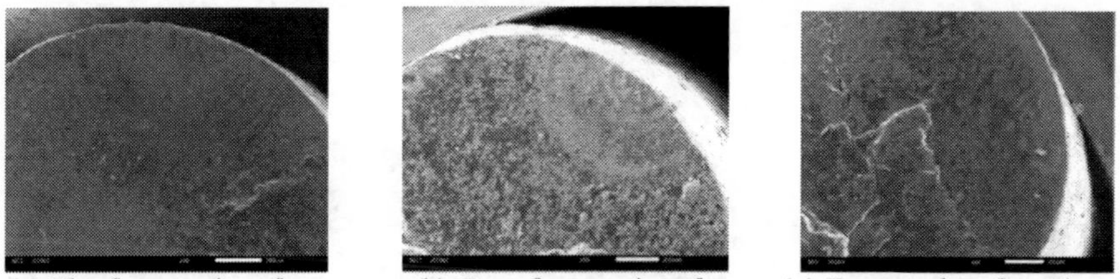

(a) Flat fractured surface (b) Torn fractured surface (c) Fractured surface showing stress concentration

Fig. 4. Typical fracture morphologies of fractured steel wires.

4.2 Surface quality of steel wires

Typical morphology of steel wire surface is shown in Fig. 5, in which a large number of microcracks exist on the surface of the steel wire. The reason for this phenomenon is that the fatigue cracks mostly originate from the surface or sub-surface of steel wires. Surface quality of steel wires has important effect on its fatigue resistance. The surface cracks would continue to extend inward and finally induce the fracture of the steel wire under applied loading stress.

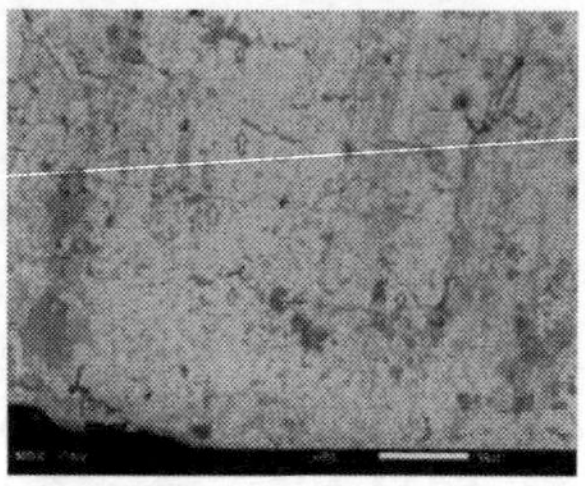

Fig. 5. Typical morphology of steel wire surface.

By observing the longitudinal section of the steel wire shown in Fig. 6, it can be seen that the coating thickness of the new steel wire is larger than that of the old one due to the wear of steel wires in service. In addition, the galvanized layers of both steel wires are uneven, which may affect the surface quality of wire ropes, and even induce localized severe wear and fatigue fracture. Therefore, reasonable measures should be adopted in the production of galvanized layers in order to improve the surface quality of wire ropes.

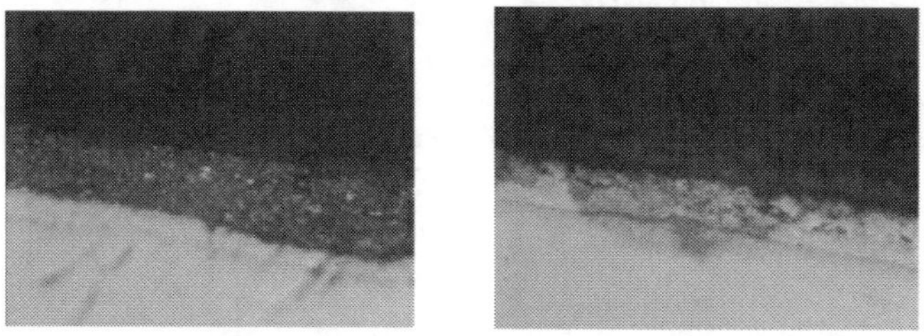

(a) a new wire rope (b) an old wire rope

Fig. 6. Optical microscopy of longitudinal cross-section of steel wires (×500).

4.3 Microstructure of steel wire

Two different areas on the fatigue fracture surface of a steel wire were chosen to conduct Energy Dispersive Spectrometer (EDS) spot scanning analysis, as shown in Fig. 7. EDS analysis shows that the carbon content of the spot closer to the surface is higher than the other, resulting in higher hardness of the steel wire surface than that of the inner area. When under alternating stress, cracks would firstly originate on the interface close to the steel wire surface between the two steel microstructures, and the fracture of steel wire finally occurs with the crack continuously extending inward.

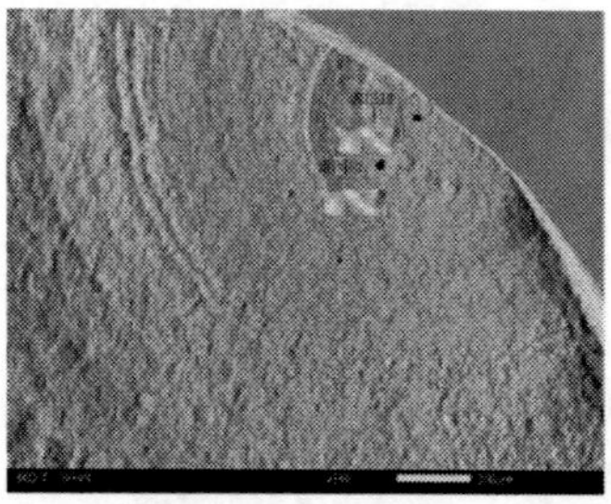

Fig. 7. Locations of EDS spots scanning on the fatigue fracture surface of a steel wire.

Metallographic structures of steel wires after being corroded by 4 wt.% of nitric acid in alcohol solvent is shown in Fig. 8. It shows that the metallographic structure of the old wire rope is slenderer than that of the new one due to the elongation of the wire rope in actual service. Work hardening phenomenon occurs to the elongated metallographic structure, leading to the increase of brittleness [12]. Scars caused by squeezing between steel wires would appear on the surface of steel wires during use. Therefore, the metallographic structure and its orientation near the scar is obviously different from that in other areas, and the fatigue cracks tend to originate in the transition region between the two microstructures. Meanwhile, the regional agglomeration of single black microstructure named cementite and single white microstructure called ferrite exist in both wire ropes, resulting in different performance among different microstructures and initial cracks. The early fatigue fracture of steel wires is strongly related to uneven microstructure [13].

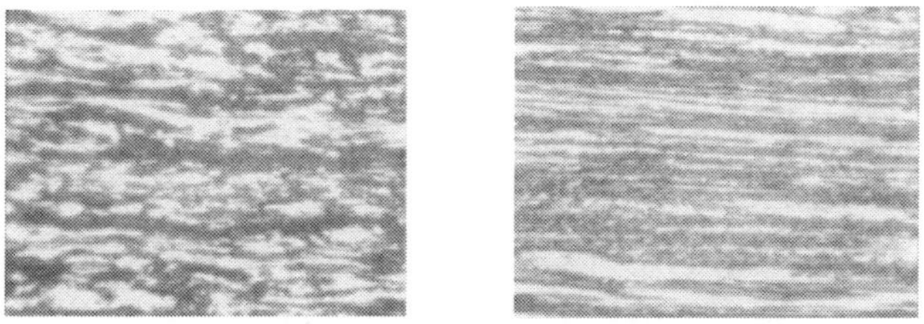

(a) a new wire rope (b) an old wire rope

Fig. 8. Metallographic structures of steel wires (×800).

4.4 Internal defects of steel wires

Internal defects of steel wires observed by a metallographic microscope are shown in Fig. 9. It can be seen that serious internal discontinuous defects such as slag inclusions or air voids exist in both new and old wire ropes. The defects would become fatigue crack sources, and therefore internal defects of steel wires should be strictly controlled and minimized in the production of wire ropes.

(a) a new wire rope (b) an old wire rope

Fig. 9. Internal defects of steel wires (×500).

5 Precautions for safe use of wire ropes

Wire ropes often bear high-speed impact loads in actual service. Based on the above analysis of failure modes, failure mechanisms and failure cases, precautions for the safe use of wire ropes in service are summarized as follows:

(1) Frequent sharp change of operating speed will induce high-speed impact loads in actual service. A wire rope may fracture when the high-speed impact loads exceed its allowable working stress. Although high-speed impact loads do not necessarily result in the fracture of wire ropes, it would certainly shorten the service life of wire ropes severely. Therefore, the sharp acceleration, deceleration and emergency braking should be avoided during the use of wire ropes.

(2) Overload operations should be strictly prohibited for wire ropes. Overload operations will obviously increase the wear degree of steel wires, and may cause the fracture of steel wires, which would greatly affect the operation security of wire ropes.

(3) Friction between wire ropes and other objects except pulley under high speed operation should be strictly prohibited, because the friction under high-speed operations

will generate instantaneous frictional heat, which may lead to the formation of martensite structure on the surface of steel wires and early cracks of steel wires.

(4) Sliding out from the corresponding pulley groove should be strictly prohibited for wire ropes, because it would bring about squeezing deformation, knotting, steel wire fracture and spiral strands fracture, which will shorten the service life of wire ropes or even cause catastrophic consequences.

(5) Disheveled twist should be strictly prohibited for wire ropes. The wire rope on the drum winding should be as orderly as possible; otherwise it will induce mutual squeezing, structural damage and early fracture of steel wires, directly affecting the service life of wire ropes.

(6) Wire ropes should be avoided from water and wet sand in actual service, and they should be used in the environment as dry as possible. Galvanized wire ropes or other anti-corrosion wire ropes are recommended to serve in severe service environment. Lubrication is also recommended during use for common wire ropes.

6 Conclusions

Wire ropes are often subjected to high-speed impact loads in a range of industrial applications. The loads applied on the wire rope have the following characteristics: alternating impacts, short-time overload, and small movements among adjacent spiral strands. We analyze the typical failure modes of wire ropes in this paper. Based on the analysis of failure modes, failure mechanisms and failure cases, the precautions for the safe use of wire ropes have been summarized. The primary precaution is that the sharp acceleration, deceleration and emergency braking should be avoided during the use of wire ropes.

References

1. J. Ma, S.R. Ge, D.K. Zhang, J. China. Univ. Min. Techno. **18**, 475 (2008).
2. O.A. Goroshko, Int. Appl. Mech. **43**, 64 (2007).
3. D.G. Wang, D.K. Zhang, S.R. Ge, Ind. Lubr. Tribol. **65**, 436 (2013).
4. D. Elata, R. Eshkenazy, M.P. Weiss, Int. J. Solids Struct. **41**,1157 (2004).
5. D.K. Zhang, S.R. Ge, D.S. Xiong, Tribol. **21**, 362 (2001).
6. Y. Shen, D.K. Zhang, D.G. Wang, Tribol. **30**, 404 (2010).
7. D.G. Wang, D.K. Zhang, W.J. Zhao, et al., Mater. Sci. Eng. A **596**, 80 (2014).
8. K. Kumar, J. Botsis, J. Appl. Mech. **68**, 432 (2001).
9. X.H. Lu, Z.P. Chen, S.G. Wang, Met. Prod. **39**, 1 (2013).
10. J.M. Han, L.Y. Wang, W.Y. Chen, J. Taiyuan Uni. Techno. **31**, 218 (2000).
11. F.G. Liu, Met. Heat Treat. **9**, 42 (1999).
12. M. Giglio, A. Manes, Eng. Fail. Anal. **12**, 549 (2005).
13. K. Kumar, J.E. Cochran. J. Appl. Mech. **54**, 898(1987).

Technics Research on Polycrystalline Cubic Boron Nitride Cutting Tools Dry Turning Ti-6AL-4V Alloy Based on Orthogonal Experimental Design

Yunhai Jia[1,2]*, *Lixin* Zhu[2]

[1]Beijing Institute of Electro-machining, 100191 Beijing, China
[2]Beijing Key Laboratory of Electrical Discharge Machining Technology, 100191 Beijing, China

Abstract. Ti-6Al-4V components are the most widely used titanium alloy products not only in the aerospace industry, but also for bio-medical applications. The machine-ability of titanium alloys is impaired by their high temperature chemical reactivity, low thermal conductivity and low modulus of elasticity. Polycrystalline cubic boron nitride represents a substitute tool material for turning titanium alloys due to its high hardness, wear resistance, thermal stability and hot red hardness. For determination of suitable cutting parameters in dry turning Ti-6AL-4V alloy by Polycrystalline cubic boron nitride cutting tools, the samples, 300mm in length and 100mm in diameter, were dry machined in a lathe. The turning suitable parameters, such as cutting speed, feed rate and cut depth were determined according to workpieces surface roughness and tools flank wear based on orthogonal experimental design. The experiment showed that the cutting speed in the range of 160~180 m/min, the feed rate is 0.15 mm/rev and the depth of cut is 0.20mm, ideal workpiece surface roughness and little cutting tools flank wear can be obtained.

1 Introduction

Titanium alloys find wide application in many industries. Titanium after being alloyed with aluminum, vanadium and other elements is highly suitable to be used in aircraft, naval ships, armor plating, missiles and aircrafts, due to their unrivalled and unique combination of high strength-to-weight ratio and high resistance to corrosion. Most importantly, it resists the crack growth and creep elongation even at high temperatures. Ti-6Al-4V is the mostly used alloy in aircraft applications, almost 50% of all alloys [1,2]. The machinability of titanium alloys is impaired by their high temperature chemical reactivity, low thermal conductivity and low modulus of elasticity [3]. The machine-ability of Ti-6Al-4V has been investigated actively worldwide since 1980s and most researchers have found that polycrystalline cubic boron nitride (PcBN) and polycrystalline diamond (PCD) tools can be used to machine Ti-6Al-4V effectively [4]. The development of new tool materials in the past ten years, such as

* Corresponding author: jyh308401@sina.com.cn

Polycrystalline cubic Boron Nitride (PcBN), makes precision hard turning possible, in which the hardness of the workpiece is up to 58–62 HRC[5]. Parts with hardness exceeding 45 HRC can be machined by hard turning, which provides surface roughness, dimensional and shape tolerances similar to those achieved in grinding. The grinding process, of surface finishing, is now replaced by the hard turning as it owns some of the benefits [6], i.e., capability to produce, complex parts, reduced cycle time, low cost, higher productivity, etc. PcBN (polycrystalline cubic boron nitride) is the material most commonly used in tools for hardened steel turning applications due to its high hardness, wear resistance and thermal stability [5].

There have been many valuable research results from Ti-6AL-4V alloy machining. Bouacha et al. optimized hard turning operation in respect of cutting speed, feed rate, depth of cut and machining time. The tool wear, surface roughness, cutting forces, and metal volume removed were the responses. The cutting speed and feed rate were the most significant factors for surface roughness; specifically, the higher cutting speed produced a better surface finish [7]. In DA Fadare study, the steepest descent method was used to study the effects of cutting parameters on surface roughness of machined Ti-6Al-4V alloy workpiece at high-speed conditions. The results revealed that surface roughness was more sensitive to variation in feed rate followed by cutting speed and depth of cut [8]. SK Goyal et al analyze the performance of precision turning using conventional lathe on Ti6Al4V under dry working conditions. Various parameters that affect the machining processes were identified and a consensus was reached regarding its values [9]. In ZQ Liu study, the cutting forces and roughness for turned surfaces were measured and analyzed by using single-factor turning experiments. They pointed that PcBN tools at the cutting conditions of higher cutting speed, lower feed rate and lower depth of cut have a longer tool life, lower cutting force, better surface roughness, which means that PcBN tools are adapted in precision processes of machining the titanium alloy TC4 [10].

The aim of this investigation was to study the suitable turning parameters, such as cutting speed (v), feed rate (f) and depth of cut (a_p), which were determined according to workpieces surface roughness (W_{Ra}), average cutting tool flank wear (VB) when using PcBN cutting tools dry turning Ti-6AL-4V alloy.

2 Experimental equipment and method

Workpiece materials selected for investigation were the Ti-6AL-4V alloy rod with the composition given in Table 1. The size of the workpiece used for experimentation was round rods with dimension 100mm diameter and 300mm long. The physical and mechanical properties of the Ti-6AL-4V alloy were listed in Table 2. They were machined by Polycrystalline cubic Boron Nitride cutting tools (BZN 9100 produced by DI Corporation) in dry turning in a lathe (CAK6150). The cutting tool geometry parameters were shown in Table 3. The TR240 surface roughness tester, 19JPC-V universal tool maker's microscope and S-4800 scanning electron microscope (SEM) are selected as experimental analyzer.

Table 1. Composition of workpiece (Φ100×300 mm) material (%).

Workpiece material	C	AL	V	Si	Fe	N	H	O	Ti
Ti-6AL-4V	0.1	6.2	4	0.15	≤0.25	0.05	0.0125	≤0.2	Others

Table 2. Physical and mechanical properties of the Ti-6AL-4V alloy.

Density (g/cm^3)	Melting point (℃)	Specific heat capacity (J/Kg ℃)	Thermal conductivity (W/mK)	Electric resistivity (ohm.cm)	Fracture Toughness ($MPam^{1/2}$)	Elastic modulus (GPa)	Hardness (HRC)
4.43	1649	526.3	6.7	1.78×10^{-4}	75	113.8	38

Table 3. Cutting tool geometry parameters.

Rake angle γ_0(°)	Flank angle α_0(°)	Blade angle λ_s(°)	Edge angle k_r(°)	Nose radial r_ε (mm)	Vice chamfering width b_{r1}(mm)	Vice chamfering angle γ_{01}(°)
3	8	0	80	0.5	0.2	10

Surface roughness is used as the critical quality indicator for the machined surface. Formation of a rough surface is a complicated mechanism involving many parameters. The quality of the workpiece (either roughness or dimension) is greatly influenced by the cutting conditions, tool geometry, tool material; machining process; workpiece material and tool wear during cutting.

Orthogonal experimental design has been used to analyze relationships between the turning parameters and workpiece surface quality. The combinations of the factors at their different levels and the corresponding measured values of the surface roughness generated after the machining operations are given in Table 4.

Table 4. Orthogonal experimental design and experiment results (continue turning 2 minute).

S/N	Cutting speed v (m/min)	Feed rate f (mm/rev)	Depth of cut a_p (mm)	Workpiece surface roughness W_{Ra} (um)	Flank face wear VB (mm)
1	120	0.10	0.15	0.45	0.012
2	120	0.15	0.20	0.49	0.009
3	120	0.20	0.25	0.95	0.090
4	160	0.10	0.20	0.39	0.011
5	160	0.15	0.25	0.45	0.019
6	160	0.20	0.15	0.47	0.029
7	200	0.10	0.25	0.60	0.035
8	200	0.15	0.15	0.41	0.026
9	200	0.20	0.20	0.46	0.085

3 Study on relationship between turning parameters and workpiece surface roughness

The values of workpiece surface roughness were compared in Fig. 1(a) on the condition of the feed rate of 0.15 mm/rev, cutting speeds from 80 to 220 m/min, three different depths of cut 0.15, 0.20 and 0.25mm. From this figure, it can be known that with increasing of cutting speed, workpiece surface roughness values decreased until a minimum value was reached, after that surface roughness values increased. The lowest average value of workpiece surface roughness got obtained at 180 m/min cutting speed. The values of workpiece surface roughness were compared in Fig. 1(b) in the condition of depth of cut 0.20, feed rate from 0.05 to 0.40 mm/rev, three different cutting speeds 120, 160 and 200 m/min.

From this figure, it can be known that with increasing feed rate, workpiece surface roughness values increase. Fig. 1(c) shows the relationship between depths of cut with workpiece surface roughness. It can be seen in this figure that workpiece surface roughness with the increase of cutting depth, slightly reduced then increased, because of the low elastic modulus of titanium alloy, so when the cutting depth is less than 0.05mm, processing produced the spring-back deformation, so that the processing quality rather than cut deep 0.15mm. With the further increase of the cutting depth, the dynamic cutting force also increases, which leads to the deformation of the titanium alloy and the vibration of the machining system, and the surface roughness increases.

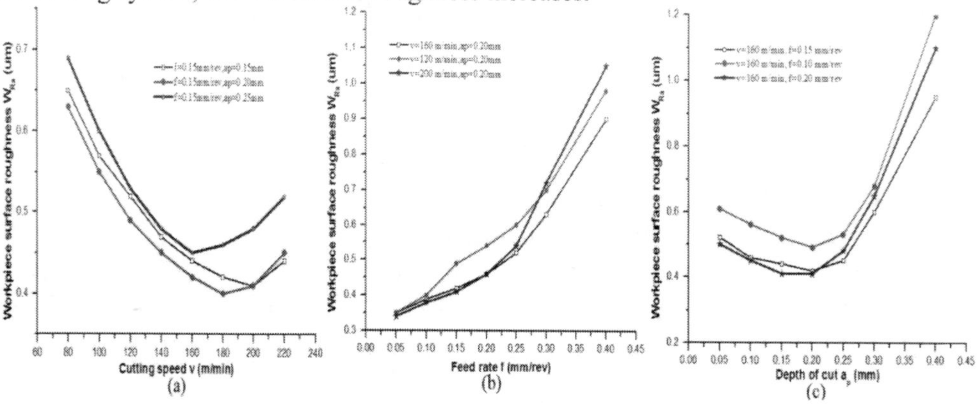

Fig. 1. Relation curve between turning parameters and workpiece surface roughness.

4 Research on cutting tool's flank wear

Fig. 2(a) shows the worn surface of PcBN cutting tools rake face after using PcBN continue dry turning Ti-6AL-4V alloy rod ten minutes. Fig. 2(b) shows the worn surface of PcBN cutting tools flank face after using PcBN continue dry turning Ti-6AL-4V alloy rod ten minutes.

Fig. 2. SEM of PcBN cutting tools face.

The effects of cutting speed and feed rate values on the cutting tools flank wear were given in Fig. 3(a) and (b). From Fig. 3(a), it can be seen that flank face wear slow vary with the cutting speed increase when cutting speed is below 180 m/min, after that flank wear huge increase. From Fig. 3(b), it can be seen that flank face wear slow increase in the feed rate increase when the feed rate is below 0.15 mm/rev, after that flank wear huge increase.

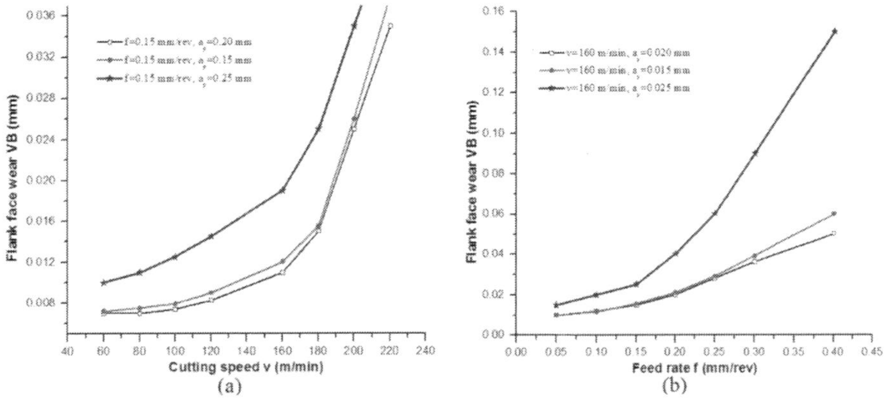

Fig. 3. Relation curve between turning parameters and cutting tools flank wear.

5 Conclusions

Through the test and analysis of the above, the conclusion can be obtained as follow:

(1) The surface roughness was more sensitive to variation in cutting speed than the depth of cut when the feed rate is below 0.20 mm/rev.

(2) The effect of feed rate and depth of cut on flank wear was higher than that of cutting speed during PcBN cutting tools dry turning Ti-6Al-4V alloy workpiece.

(3) Good surface quality can be achieved at low feed rate and depth of cut with high cutting speed in using PcBN cutting tools dry turning Ti-6Al-4V alloy. It can be obtained ideal workpiece surface roughness and little cutting tools flank face wear, the cutting speed in the range of 160–180 m/min, the feed rate is 0.15 mm/rev and the depth of cut is 0.20mm.

This lesson is supported by Beijing Natural Science Foundation the Grant No. 3162013 and Beijing Academy of Science and Technology Innovation Team Project No.IG201504N. The authors would also like to thank the anonymous reviewers whose comments greatly helped in making this paper better organized and more presentable.

References

1. V. A. Joshi, *Titanium Alloys: An Atlas of Structures and Fracture Features* (CRC Press, London, 2006)
2. Ezugwu E, Bonney J, Yamane Y., Journal of Materials Processing Technology **134**, 3 (2003)
3. Rahman M, Wang Z-G, Wong Y-S, JSME Int J **49**, 11 (2006)
4. Adilson Jose de Oliveira, Anselmo Eduardo Diniz, Davi Janini Ursolino., Journal of Materials Processing Technology **209**, 6 (2009)
5. R Shetty, TK Jose, GD Revankar., International Journal of Current Engineering and Technology **4**, 6 (2014)
6. Aouici H, Yallese MA, Chaoui K., Mabrouki T, Measurement **16**, 3 (2012)
7. Bouacha K, Yallese MA, Khamel S., Belhadi S, Int J Refract Met Hard Mater **45**, 6 (2014)

8. DA Fadare, WF Sales, EO Ezugwu, J Bonney, Journal of Applied Sciences Research **16**, 5 (2009)
9. SK Goyal, Vinayagamoorthy, International Journal of Current Research **36**, 11 (2012)
10. ZQ Liu, AI Xing, LI TT, Y Wan, Journal of Shandong University **39**, 6 (2009)

Properties and Preparation of High Entropy Alloys

Xiang Yin[1] and *Shuqiong* Xu[1*]

[1]School of Mechanical and Vehicle Engineering, Linyi University, Shandong, China

Abstract. As a consequence of multi-components, the high entropy alloys embodied serious cocktail party effect and lattice distortion. So high entropy alloys have high strength and hardness possess many comprehensive properties such as thermostability and corrosion resistance. Because of excellent mechanical properties and chemical properties, high entropy alloys have immeasurable potential of development. This paper mainly introduces the properties, preparations and applications of high entropy alloys, and finally summarizes them.

1 Introduction

From the traditional concepts' perspective, excessive components will give rise to degradation of integrated performance. After the proposal of Yeh. J .W [1]in 2004 there are so many excellent researchers carrying out extensive researches and explorations . A lot of system have been developed, such as AlxFeCoNiCrTi[2]and AlTiFeNiCuCrx[3], mainly composed of transition elements, which possesses simple crystal structures. Till now, some high entropy alloys has already received a range of generalizations and applications for the cheap costs of development.

2 The definition and crystal structures

2.1 Definition

According to the theory of Yeh. W. L, high entropy alloys generally consist of more than five main elements. The proportions between principal components are mole ratio or close to the mole ratio. And the contention of main elements is in the range of 5%-35%, while the content of minor elements is less than 5%.

2.2 Crystal Structures

In the traditional Gibbs phase rule, the number of equilibrium phases equals to N+1 in the system of alloys with N kinds of main elements. The number of high entropy alloys is not

* Corresponding author: xushuqiong11@sina.com

less than 5%. In this concept, high entropy alloys are bound to produce a wide variety of intermetallic compounds, which increased great extent brittleness and reduced the mechanical properties. Fortunately, instead of intricate intermetallic compounds, high entropy alloys formed some simple and common crystal structures such as BCC, FCC, HCP, even the non-crystalline structures [4]. This phenomenon is called high entropy effect [5], which constrained the formation of intermetallic compounds, while forming simple crystal structures.

3 Properties

3.1 High strength and high hardness.

On the one hand, the alloy will have an intensive lattice distortion effect as a result of uneven atomic radius of various elements. This kind of effect can hinder dislocation movement, which leads to very high strength and hardness. Fig.1. is a schematic diagram of the lattice distortion of high entropy alloy containing Cr. On the other hand, the "cocktail effect"[6] of high entropy alloys intensified the interaction of various alloying elements. Consequently, the high entropy alloy can reflect the comprehensive properties of all kinds of alloy elements. Therefore, the high entropy alloy reflects composite effect improving the strength and hardness.

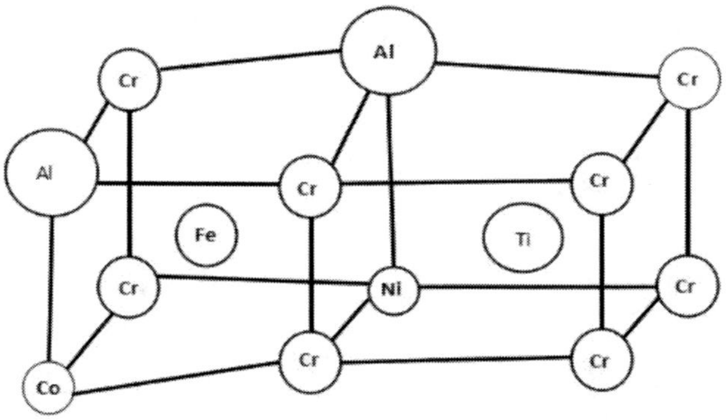

Fig. 1. Schematic diagram of lattice distortion of high entropy alloy containing Cr

As is shown in many experiments, high entropy alloys have higher high hardness. Guo Jianbin[7] melted out of four groups of high entropy alloy, ZrTiHfVNb, ZrTiHfNbMo, ZrTiHfVTa and TaMoWNbZr, and he carried out a compression experiment and measured the values of yield strength, which are respectively 1152MPa, 1134MPa, and 1089MPa. However, with the increases in the experimental load, the three kinds of alloys were soon broken, and the measured plastic strain is lower. The value of yield strength of ZrTiHfVNb is only 836MPa, but the alloy showed excellent plasticity and the plastic strain reached 55%.

3.2 Corrosion resistance

For one thing, because the crystal structure of high entropy alloy tends to be simplified, the corrosion resistance of high entropy alloys is superior to that of plenty of widely used stainless steel. For another thing, lots of high entropy alloy are dendritic segregation structures[8], which are composed of simple surface structure and body structure. And there are many disordered structures and nanometer particles in dendritic segregation structures.

The appearance of these structures has greatly increased the corrosion resistance of high entropy alloys. For example, high entropy alloy Cu0.5NiAlCoCrFeSi at room temperature exhibited better corrosion resistance than 304 and stainless steel in 1mol/L NaCl and 0. 5mol/L H_2SO_4 solutions[9].

3.3 Thermal stability

Entropy is the physical quantity of the system chaos in thermodynamics. The greater entropy means the greater the degree of chaos. According to the function of Gibbs free energy, the entropy of the system (S), temperature (T), enthalpy (H) satisfied the following mathematical relations:

$$G=H-TS \tag{1}$$

It can be known that higher entropy value makes the system more stable. And the high temperature can also reduce the free energy of the system, so the high entropy alloys have good thermal stability.

4 The preparation

4.1 Magnetron sputtering.

Magnetron sputtering[10] is to use certain high energy particle to bombard the surface of a specific material. The process is shown in Fig. 2., argon is placed in a magnetron sputtering device. Under the action of strong electric field, argon wills ionization argon ion and electrons. Then, ions are accelerated toward the cathode in the electric field to bombard target surface with high energy, resulting the occurrence of sputtering target. Therefore, placed in a magnetic field and an electric field with orthogonal distribution, the target material will emit secondary electrons, in which case, the target material will ionize more ions to continue to bombard the target; the target will be sputter deposition to the substrate surface to form a thin film.

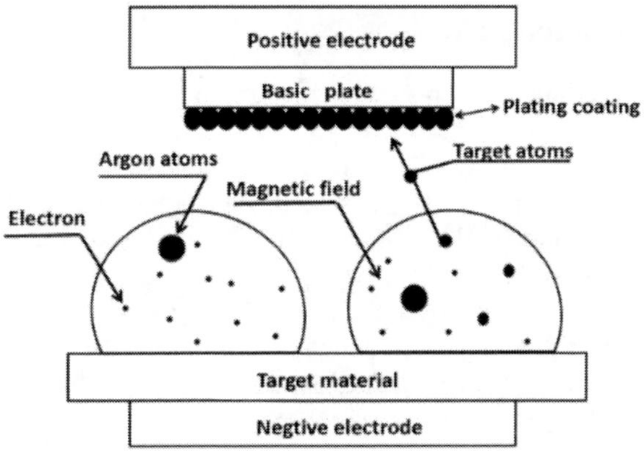

Fig. 2. Magnetron sputtering illustrative diagram

The FeCrCoNiMn high entropy alloy thin films were deposited on 304 stainless steel substrates and Si (100) substrates by magnetron sputtering at Lanzhou University[11]. The results show that the films are uniform and amorphous in a certain sputtering power. But when the sputtering power reaches a certain value, it will get a very rough film and the surface density decreases. With the increasing of sputtering power, the thin films will become smaller, and the amorphous structure will also be transformed into crystal structure. Therefore, the sputtering power can directly affect the properties of the film.

4.2 Vacuum arc melting.

Here are the brief descriptions of the process of Vacuum arc melting. Firstly, a certain proportion of the pure metal and then placed in the pot, in the vacuum furnace and the air out of the repeated charge of protective gas. Then, the arc discharge is formed between the electrode with the high current and the pot, so that the temperature can be raised so high in a short time, which can make sure the alloy, can be uniformly melted in a short time. And the final step is water cooled casting. The advantages of the method are that the refined alloy has high purity, can prevent pollution, and can refine the metal with high melting point[12].

With Vacuum arc melting, Cu0, 5AlFeNiCrTi and CuAlFeNiCrTi are melted by Gao Peng[13]. The microstructure, crystal structure, mechanical properties and corrosion resistance of the two alloys were analyzed by a series of experiments. It was found that these two kinds of alloy have a simple crystal structure with excellent mechanical properties and corrosion resistance.

4.3 Laser cladding

Laser cladding is a new technology of smelting high entropy alloys in recent years. And it broke through the dimensional limitation of traditional high entropy alloys. And the micro structure of alloys is uniform and compact with fewer micro defects [14].

Qiu Xingwu chose the matrix material of Q235 steel, coated with high entropy alloy Al2CrFeCoCuNixTi (x=0, 0. 5, 1, 1. 5, 2) with different nickel contents. He found out that the structure of these alloys are simple BCC+FCC through the XRD[15].

5 Applications

5.1 Application of cutting tools.

Comparing the friction and wear behavior of high entropy coating tool and ordinary high speed steel[16]. When the external load and rotational speed are fixed, the friction of the high entropy alloy coated cutting tool is always lower than that of the ordinary high speed steel. When the tool is rotating at high speed, the higher the friction coefficient is, the higher the friction resistance will be. At this time the chip is very easy to stick knife, and it will scratch the surface of the material, which will reduce the surface finish of the material. Accordingly, the high entropy alloy coating on the surface of the cutting tool can reduce the friction force in the cutting process, which will prolong the tool life and improve the surface quality of the machined surface. Finally, because of high strength and hardness, high entropy alloy coating can be used in high speed cutting.

5.2 Application of hydrogen storage.

Many researches and experiments show that some high entropy alloys have excellent hydrogen storage function. Qiu Jianping studied the hydrogen storage properties of these three kinds of high entropy alloys, which are respectively ScTiVCrCo, ScTiVCrNi, ScTiVCrCu. The results show that the maximum hydrogen absorption capacity of the three alloys is respectively 1.76%, 2.19%, and 2. 56%. With the increase of the number of hydrogen absorption, the amount of hydrogen absorption will be reduced correspondingly [17].

5.3 Application of chemical corrosion.

Because of eminent corrosion resistance, the high entropy alloy can be applied to all kinds of corrosive working circumstances such as the bottom of ocean ship, the occasions which are prone to electrochemical reaction and various chemical vessels, etc..

5.4 High load and high impact applications.

High entropy alloys can be used to produce high quality golf hitting surfaces, high hardness steel pipe and turbine blade, which work in high load and high circumstances. High entropy alloys can also be used for national defense science and technology, such as the track of tanks, artillery shells, etc.

References

1. X.B.Liang, M.Wei, J.B.Cheng, W.Zhang, Journal of Materials Engineering, **12**,75-79(2009)
2. W.Li, G.Z.Liu, Rare Metal Materials and Engineering, **29(10)**,941-944(2009)
3. Y.Liu, M.Chen, Y.X.Li , X.Chen, Acta Metallurgica Sinica, **43(10)**,1020-1024(2007)
4. C.Z.Yao, H.X.Ma, Chinese Journal of Applied Chemistry, **10**, 1189-1194 (2011)
5. Y.Liu, M.Chen, Y.X.Li, Materials Review, **20(4)**,4-6(2006)
6. C.F.Zhao,Heat treatment, microstructure and properties of some selected multi-principal elements high-entropy alloys. Fuzhou University(2014)
7. J.B.Guo, Study on microstructure and mechanical properties of five-element refractory high entropy alloys. Beijing Institute of Technology(2015)
8. Z.Y.Sun, Preparation of High Entropy Alloy and Research on Microstructure and Mechanical Properties. Jilin university (2014)
9. J.H.Pi, X.M.Yang, Advanced Materials Industry, **09**,71-74(2009)
10. Z.Y.Zheng, Microstructure and Properties of High Entropy Alloy Films Prepared by Direct Current Magnetron Sputtering. Jilin university(2016)
11. Y.Y.Shi, The deposition and the corrosion performance of Fe CrCoNiMnCo high entropy alloy thin films. Lanzhou University of Technology(2016)
12. X.N.Yang, W.L.Deng, X.B.Huang, L.H.Tian, Hot Working Technology, **22**,30-33(2014)
13. P.Gao, H.F.Sun, Z.L.Li, Y.Z.Yan, F.C.Li, H.Y.Zhu, Physical Testing and Chemical Analysis(Physical Testing), **11**,677-681(2010)
14. X.Y.Li, H.M.Zhu, Hot Working Technology, **22**,30-33(2014)
15. X.W.Qiu, Laser & Optoelectronics Progress,**05**,1-12(2017)
16. A.R.Zhang, Li Ye, China Surface Engineering, **04**,27-31(2013)
17. J.P.Qiu, Y.C.Luo, Y.L.Liu, K.J.Yang, Hot Working Technology, **02**,85-88(2017)

MATEC Web of Conferences **142**, 03004 (2018) https://doi.org/10.1051/matecconf/201814203004
ICMAE2017

Effects of Ar and He on Microstructures and Properties of Laser Welded 800MPa TRIP Steel

Wen-Quan Wang[1, *], *Shu-Cheng* Dong[1], *Fan* Jiang [1], and *Ming* Cao [1]

[1]School of Material Science and Engineering, Jilin University, Changchun, PR China

Abstract. Fiber laser welding of cold rolled TRIP steel (transformation Induced Plasticity steel) sheet with tensile strength of 820MPa and thickness of 1.4mm was carried out using shielding gases Ar and He, respectively. For the same laser power and welding speed, the effects of different shielding gases on penetration and bead section morphologies were investigated. The microstructures and properties of the TRIP steel joints were also studied. The investigation showed that higher penetration and lower porosity could be obtained under shielding gas He using the same laser power and welding speed. The microstructures of the TRIP joint mainly included martensite and retained austenite. But the joint microhardness and tensile strength were higher under the shielding gas He. The tensile strength of the welded joint perpendicular to the weld line was equal to that of the base metal. But the tensile strength of the joint parallel with the weld line was higher than that of the base metal. The plasticity and formability of the welded joint were impaired due to the formation of martensite in the weld metal.

1 Introduction

Weight reduction of automobiles is one of the effective measures for decreasing gas consumption and minimizing environmental pollution on the premise that safety provisions are guaranteed. It was estimated that 10% of weight reduction generally may bring about 5% less of gas consumption [1]. Therefore, these years light metal materials such as Al alloys and Mg alloys have been increasingly used in automobile industry. But the high cost of manufacturing all light metal car body has greatly limited the usage expansion. During the past years, ultra-high strength steels (UHSS) such as transformation induced plasticity (TRIP) steels and dual phase (DP) steels were increasingly applied in automobile industry due to their good combination of high strength and formability. The excellent mechanical properties of TRIP steel mainly contribute to the ferrite matrix, for ductility, bainite, for strength, and retained austenite, for uniform elongation produced by martensite transformation from austenite when subjected to external tensile stress. For the sake of the superior properties of TRIP steel, thinner steel sheets can be adopted for the same strength

* Corresponding author: wwq@jlu.edu.cn

requirement to reduce the vehicle weight but improve the vehicle performance at the same time [2-5].

Recently, laser cutting and welding technology has been increasingly used in automobile and its components supplier industries [6-8]. Laser beam welding (LBW) is considered as an optimal joining alternative for its high quality, high productivity and low thermal deformation. In this paper, microstructural characteristics and mechanical properties of fiber laser welded TRIP steel sheet with tensile strength of 820MPa were investigated to provide some fundamental knowledge about laser welding TRIP steels for the engineers in automobile industry.

2 Experiment

In this work cold rolled TRIP steel sheet with tensile strength of 820MPa and thickness of 1.4mm was used as base metal. Its chemical compositions and microstructures were shown in Table 1 and Fig.1. The study showed the multi-phase microstructures of the base metal including ferrite matrix plus bainite and a little retained austenite. This was attributed to the heat treatment history of the base metal. Generally, transformation induced plasticity steel was produced by intercritical annealing followed by austempering at the bainite transformation temperature range.

Table 1. The chemical compositions of base metal (wt. %)

C	Mn	Si	S	P	Fe
0.09	1.7	1.0	0.005	0.06	balance

Laser beam welding was carried out with fiber laser system (IPG YLS 4000) under different shielding gases Ar and He, respectively. The focus distance was 180mm and beam size was 0.4mm in diameter. The micro Vickers hardness was measured near fusion zone with a load holding time of 12sec and a load of 3.0N. The specimens for metallographic analysis were prepared following the steps of cutting, grinding, polishing and etching. The microstructures of the weld zone were investigated with scanning electron microscopy (SEM). The phase's identification was determined by X-ray diffraction (XRD). Tensile tests were conducted under load perpendicular to the weld line and load parallel with the weld line at a crosshead moving speed of 4.5mm/min.

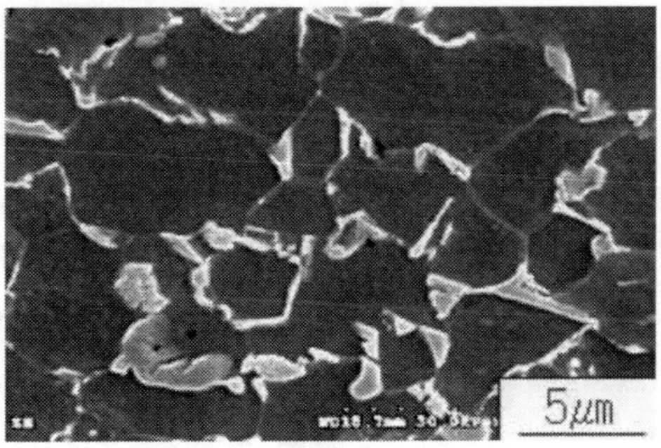

Fig. 1. Microstructures of the base metal

3 Results and discussions

3.1 Effects of shielding gases Ar and He on laser welding of TRIP steel

Fig.2 revealed that the relationship between laser power and welding speed for full penetration under different shielding gases Ar and He. Obviously, with the rise of laser power, the welding speed increased for full penetration. But for the same laser power, higher welding speed can be used under shielding gas He compared with shielding gas Ar for full penetration. This should be attributed to the higher ionization energy of He, which may lead to the increase of laser beam energy density

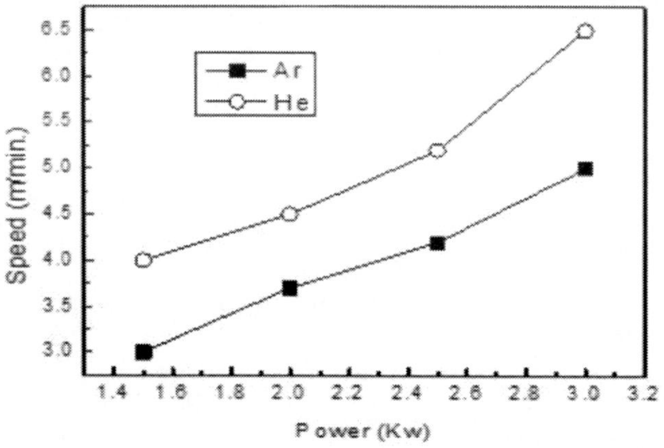

Fig. 2. Relationship of laser power and welding speed for full penetration under Ar and He

In order to understand the effects of shielding gases on bead morphologies, the macrostructure of beads at different welding speed was observed. Fig.3 presented that bead and bead cross-section morphologies at the same welding speed of 4m/min. under shielding gases Ar and He, respectively. The experiments indicated that sound beads could be obtained using proper welding parameters under shielding gases of either Ar or He, as shown in Fig3(a). No obvious difference for bead surface was detected under different shielding gases. But the porosity rate was higher for the TRIP joint under shielding gas Ar than that under He. For fiber laser welding of TRIP steel, the weld bead cross-section of full penetration joint presented drum shape with wide upper and bottom part but narrow middle part. For the same welding speed and laser power, the cross-section of bead under shielding gas He had bigger size compared with that under shielding gas Ar, as shown in Fig.3 (b) and Fig.3 (c). This was associated with high energy density of laser beam in shielding as He. The study also showed that the weld bead cross-section for partial penetration would become trapezoid shape with narrow upper part and wide bottom part with the increase of welding speed or decrease of the laser power, as shown in Fig.4 (b). The further investigation demonstrated that the scope of fusion zone (FZ) and heat affected zone (HAZ) of the TRIP joint would shrink at the same time (Fig.4 (a)).

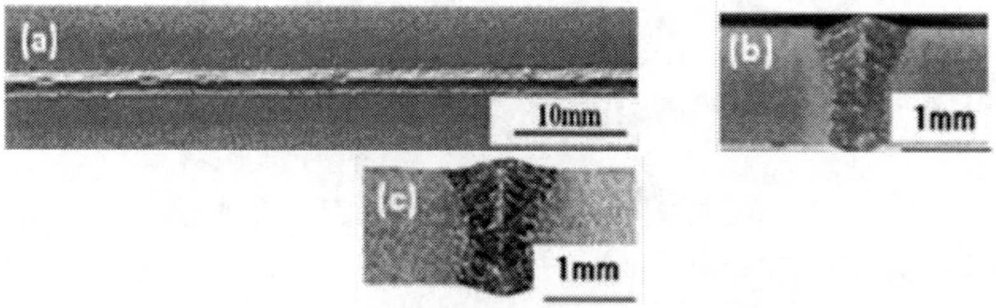

Fig. 3. Morphologies of weld bead and bead cross-section under welding speed of 4.0m/min., shielding gases Ar and He (a) weld bead surface, (b) bead cross-section under Ar, (c) bead cross-section under He

(a) Diagram of bead cross-section (b) 6.0m/min

Fig. 4. Morphologies of bead cross-section at different welding speed

Fig.5 showed the microstructures of different positions at the welded joint at welding speed of 4/min. At the position of weld metal and HAZ near weld metal, the amount of martensite increased rapidly because of high cooling speed, which could contribute to the austenite transformation to martensite, as shown in Fig.5 (a). At the position of HAZ center (Fig.5 (b)), the microstructures included ferrite matrix distributed with bainite and retained austenite. A little martensite could also be detected due to the high heating and cooling speed.

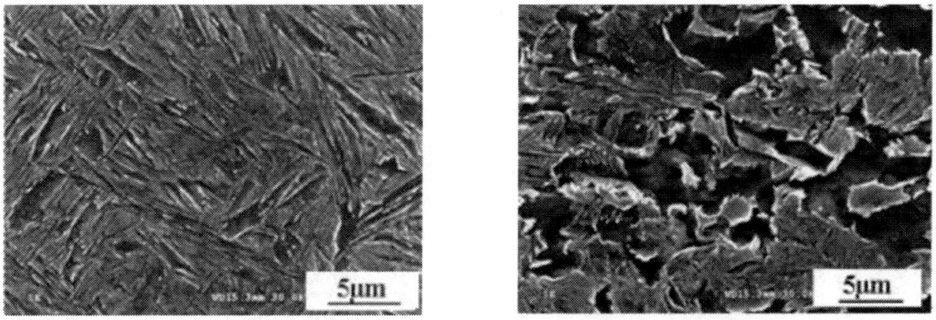

(a) Microstructure of weld metal (b) Microstructure of HAZ center

Fig. 5. Microstructures of different welded joint positions (4m/min)

3.2 Mechanical properties of laser welded TRIP steel joint.

For ultra-high strength steel (UHSS) sheets applied in automobile industry, the hardness of materials is one of critical factors which determine the formability during the process of pressing or stamping. Fig. 6 showed the hardness distribution of welded joint prepared under different shielding gases Ar and He, respectively.

The weld metal and HAZ had maximum hardness, which decreased rapidly approaching the base metal. This hardness distribution could be attributed to the microstructural constitution of various positions at the weld zone. As discussed above, the amount of martensite gradually decreased from the central welding line to the base metal at weld zone. Therefore, the maximum hardness occurred at the fusion zone or HAZ near the fusion zone. In addition, Fig.6 revealed that the maximum hardness at the fusion zone was higher under the condition shielding gas He. This was attributed to the lower porosity rate under shielding gas He. Further, Fig. 6 also indicated that the fusion zone was wider for the same power and welding speed under shielding gas He.

Fig. 7 illustrated the tensile test results under a load parallel with the weld line of TRIP joint prepared under shielding gases Ar and He, respectively. The test showed that both the yield strength and tensile strength of the specimens were a little higher than those of the base metal. The fracture always occurred along the plane vertical to the tensile direction and the fracture surface was like drum shape, with a narrow upper and lower part but a wider middle part, as shown in Fig.8. This meant that the weld metal had different elongation from that of the base metal. In other words, the weld metal has lower elongation than that of the base metal. Further study revealed that the tensile strength of the joint prepared under shielding gas He was higher than that of the joint prepared under Ar. But the yield strength of two kinds of joints had little difference. The analysis to the fracture of the joint indicated that could be explained by higher hardness and lower porosity of the joint prepared under He.

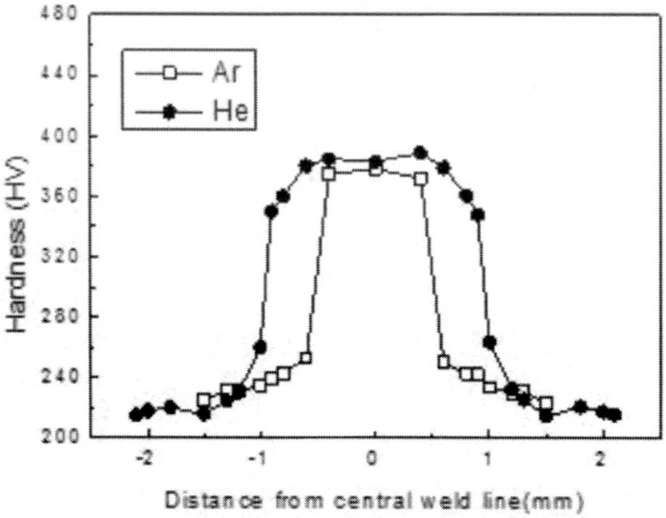

Fig. 6. Hardness distribution at weld zone

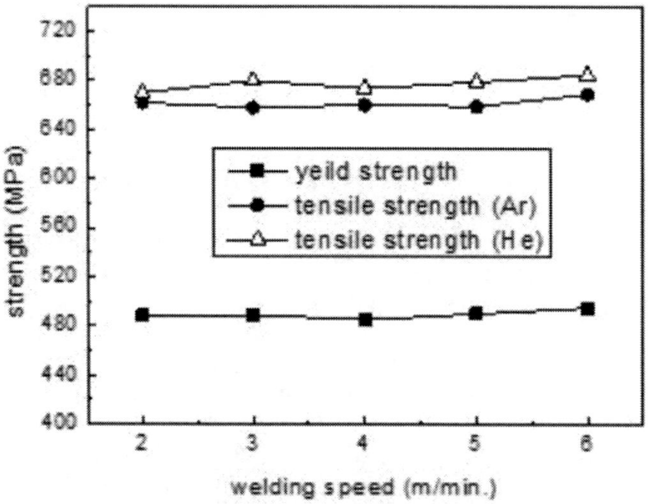

Fig.7. Tensile test of laser welded TRIP joints

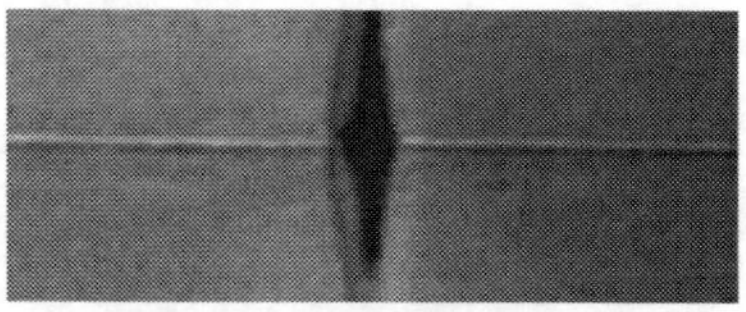

Fig. 8. Laser welded TRIP joint specimen after tensile test

4 Conclusions

(1) Shielding gases He and Ar have different effects on laser welding of TRIP steel sheet. For the same laser power, higher welding speed could be obtained under shielding gas He than that under shielding gas Ar in the case of full penetration. The joint prepared under shielding gas He had lower porosity than that prepared under Ar.

(2) The weld metal of joint mainly consisted of martensite due to high cooling speed. HAZ consisted of martensite, ferrite, bainite and retained austenite. Therefore, the weld metal or heat affected zone near the weld metal had maximum hardness for the whole welded joint.

(3) Both the yield strength and the tensile strength of the welded joint tested parallel with the weld line were a little higher than those of the base metal. The tensile strength of the TRIP joint prepared under shielding gas He was higher than that of the joint prepared under shielding gas Ar. Laser welding decreased the plasticity of TRIP steel joint due to the formation of hard phases.

References

1. H.X. Yin, A. Zhao, X. Li, Int. J. Min. Met. Mater, **22**, 262 (2015)
2. A. Grajcar, M. Różański, S. Stano, A. Kowalski, J. Mater. Eng. Perform, **23**, 3400 (2014)
3. O. Kwon, K.Y. Lee, G.S. Kim, et al, Mater. Sci. Forum, **638-642**, 136 (2010)
4. X.Y. Hou, Y.Q. Wang, R.C. Xu, et al, S. J. R, **4**, 156-161. (2014)
5. C. Seang, A.K. David, E. Ragneau, Physcs. Proc, **41**, 448-459 (2013)
6. L. Romoli, C.A. Rashed, Int. J. Adv. Manuf. Tech, **81**, 563 (2015)
7. X.L. Gao, L.J. Zhang, J. Liu, et al, Mat. Sci. Eng. A-Struct, **559**, 14 (2013)
8. H. He, S. Lin, C. Yang, et al, J. Mater. Eng. Perform, **22**, 3315 (2013)

Evaluation of Cyclic Durability in Surface-modified Layers with TiNiZr Thermoelastic Phase Transformations

Peter Rusinov[1] *, and *Zhesfina* Blednova[1]

[1]Kuban State Technological University, Engineering and Service Centers, Dynamics and Strength of Machines, Moskovskaya 2, Krasnodar, Russia

Abstract. This paper estimated longevity and damaging of surface materials modified by alloys with thermoelastic phase transformations, taking into account: the energy consumed for the formation of failure micro-surface in one cycle, in terms of the shape of the hysteresis loop; energy intensity of the alloy, in the light of phase composition by the results of X-ray diffraction analysis. The energy intensity and theoretical strength of ternary TiNiZr systems are calculated and plotted according to thermodynamic characteristics and phase diagrams of ternary systems. This paper carried out tests of steels with surface-modified TiNiZr layers under conditions of multicyclic fatigue and frictional friction.

1 Introduction

Two approaches are used to describe phenomena occurring in alloys with thermoelastic phase transformations: a straight one, when the system evolution laws are described by the consideration of transformation micromechanisms [1, 2], or directly through experiments [3]; a thermodynamic one, which is based on the analysis of the free system energy and includes the state variables that allow determining the volume fractions of various martensite forms [4]. To develop a methodology for predicting the fatigue strength of cyclic durability, the universal approaches applicable to complex systems, including statistical ones, energy (thermodynamic) and synergistic ones, can be considered as a theoretical basis. For surface-hardened materials or composite surface-modified materials, the most suitable is the energy (thermodynamic) approach, based on the analogy of the destruction, plastic deformation and melting, and the fractal-synergetic approach associated with it. The choice of the energy criterion based on the analogy of deformation, fracture and melting, evaluation of the cyclic durability of surface-modified materials is explained by the following fact: energy (energy intensity, energy content) is an additive quantity and can be determined for multicomponent materials, taking into account the chemical composition of the surface layer responsible for the destruction. It also can be determined by thermodynamic characteristics and phase diagrams of systems. Energy intensity can be

* Corresponding author: ruspiter5@mail.ru

determined taking into account the technological operations of surface modification based on the results of X-ray diffraction analysis.

The purpose of this paper is to estimate the cyclic durability in surface-modified layers with thermoelastic phase transformations of TiNiZr.

To evaluate longevity, we proposed a version of the energy model of damage accumulation in a surface-doped material for two component alloys, proposed by the following authors [5, 6]. Models [5, 6] consist in the fact that for the destruction of material it is necessary to spend part of the magnitude of the irreversible deformation work. The accumulation of damage consumes not all, but only a fraction of the irreversible deformation work. In order to establish this fraction which depends on the conditions of heat exchange, on the degree of deformation, etc., it is necessary to refer to a physical or mechanical experiment.

The energy consumed on the distortion of the crystal lattice is determined by the formula [6]:

$$H = \int_{T}^{T_m} c_p \cdot dT \tag{1}$$

Where c_p - specific heat, J / kg·K;

T - temperature, K;

T_m - melting point, K.

The ultimate energy (energy intensity) absorbed by the body at the temperature T at its melting is determined by the formula:

$$Q = \int_{T}^{T_m} c_p \cdot dT + Lm \tag{2}$$

Where Lm - latent heat of melting.

The theoretical strength is determined by the formula [6]:

$$\sigma_{\text{теор}} = \sqrt{E \cdot Q} \tag{3}$$

Where E - the Young's modulus, MPa.

The latent heat of melting is determined from the formula [6]:

$$L_m = \Delta S_m \cdot T_m \tag{4}$$

Where ΔSm - the change in the body entropy during its melting.

The entropy of melting is found from the formula [6]:

$$\Delta Sm = A_1 \cdot \Delta S^1 + A_2 \cdot \Delta S^2 \tag{5}$$

According to the Neumann-Conne formula, the heat capacity of an alloy is calculated from the formula [6]:

$$c_p^m = A_1 \cdot c_p^1 + A_2 \cdot c_p^2 \tag{6}$$

Where c_p^1 and c_p^2 - the specific heat capacities of the alloy components.

In the present work, we calculated the energy capacity and theoretical strength of a multicomponent material with thermoelastic phase transformations of TiNiZr obtained by high-velocity oxygen-fuel spraying in a protective atmosphere (argon medium). The entropy and heat capacity were calculated for each of the phases (Fig. 1), comprising the composite layer TiNiZr, according to formulas (5) and (6). Further calculations were made using formulas (1), (2), (3), (4).

The X-ray phase analysis showed that at room temperature the initial phase state of the TiNiZr layer after the high-velocity oxygen-fuel spraying of mechanically activated powder is martensite B19phases with a monoclinic lattice, austenitic B2 phases with a cubic lattice, intermetallide phases Ni3Ti, Ti2Ni, NiZr, NiZr2, with cubic and hexagonal lattices, and we also observed a small amount of titanium oxide (TiO) – less than 2% (Fig. 1).

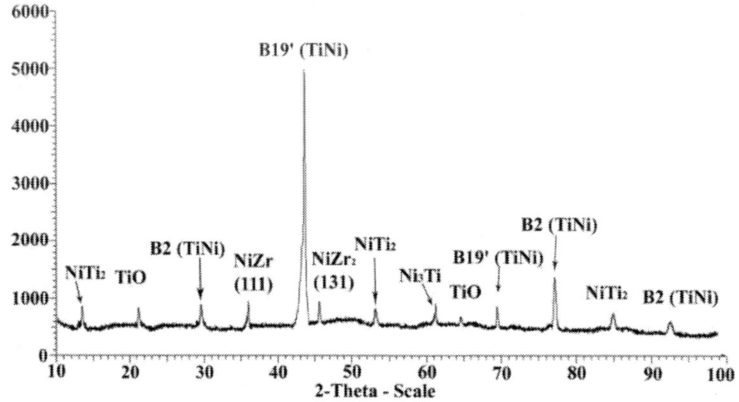

Fig. 1. X-ray phase analysis of TiNiZr alloy, after high-velocity oxygen-fuel spraying with local protection in argon.

For the phases TiNi, Ni3Ti, Ti2Ni, NiZr, NiZr2, TiO we calculated energy H, spent on the distortion of the crystal lattice, ultimate energy Q absorbed by the body at the temperature T during its melting, theoretical strength σ_{theor}. Thus, the energy H consumed on the distortion of the crystal lattice of the TiNi phase is 36.4 MJ / kg; ultimate energy Q absorbed by the body is 41.14 MJ / mol; theoretical strength of σ_{theor} is 17.91 GPa.

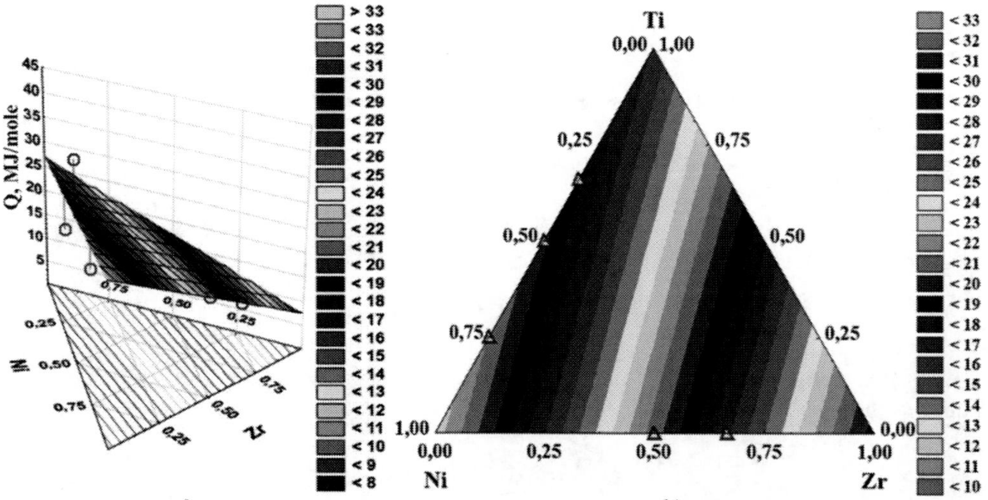

a) b)

Fig. 2. Energy intensity of Ti-Ni-Zr systems: three-dimensional (a) and two-dimensional model (b).

The calculated values of H, Q, and σ_{theor} were determined for all the component phases and the following equations were compiled on the basis of statistical processing:

- ultimate energy Q absorbed by the body at the temperature T at its melting

$$Q= 33{,}5843\cdot Ni+7{,}5379\cdot Zr+26{,}8976\cdot Ti \text{ (MJ/mole)} \quad (7)$$

- theoretical strength σ_{theor} is described by equations

$$\sigma_{theor} = 5{,}5822+1{,}0229\cdot Q-0{,}8216\cdot H \quad (8)$$

Where Ni, Zr, Ti - the percentage of elements TiNiHf (at.%).

Fig. 2 shows the energy capacity of Ti-Ni-Zr alloys (a, b). From Fig. 2 we can see that the energy intensity for the alloy of the Ti-Ni-Zr system is slightly different from the energy capacity of other compounds of this system.

Fig. 3 presents models for calculating the theoretical strength of an alloy of the Ti-Ni-Zr system.

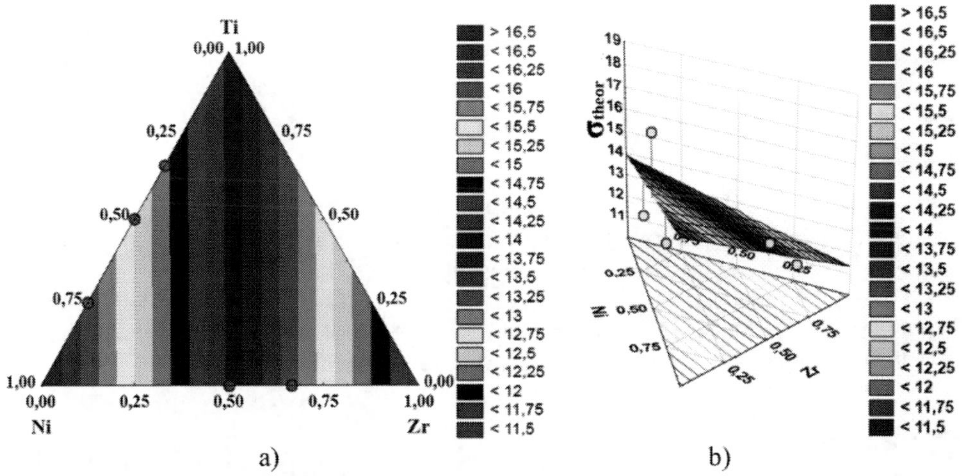

Fig. 3. Theoretical strength of Ti-Ni-Zr systems: two-dimensional (a) and three-dimensional model (b).

2 Operational properties of surface-modified TiNiZr layers

As we can see from the samples' tests (steel 1045 + TiNiZr layer) on multicyclic fatigue with bending at rotation showed, the durability increases (Fig. 4a). The endurance limit (σ_{-1}) of uncoated steel 1045 was 275 MPa, and after surface modification with a TiNiZr memory alloy it was 430 MPa (increased by 36%). In the process of testing for wear of samples with a surface layer of TiNiZr, an increase in the surface temperature in the friction zone is observed, which causes a martensitic-austenite transformation. Sufficiently large pressures arising from friction cause the plasticity of the transformation effect due to the formation of martensite stress. The combination of these processes determines the wear resistance of the sample. Experience shows that the higher the reversible deformations, the higher the wear resistance is. The increase in wear resistance of steel 1045 with a surface layer of TiNiZr after HVOF + Thermomechanical treatment is 3.5-3.6 times compared to that of steel 1045 (Fig. 4, b).

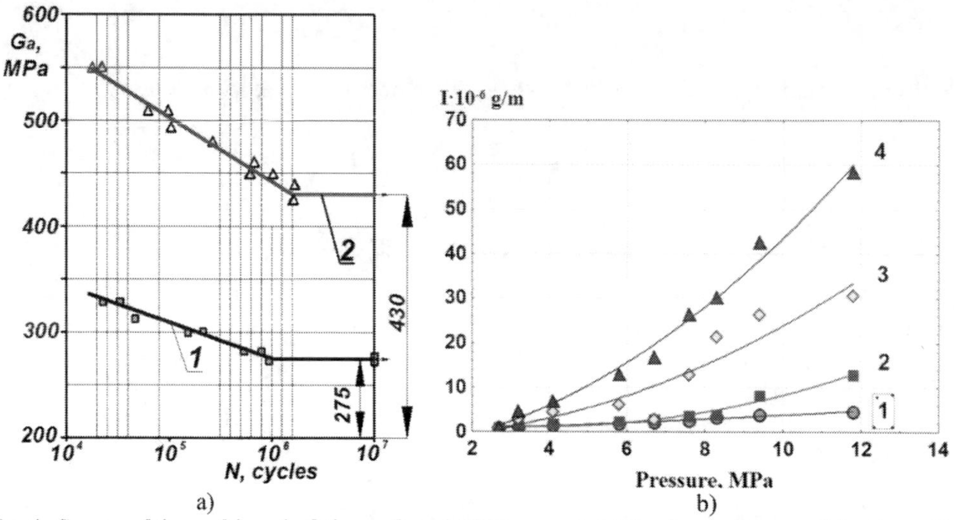

a) b)

Fig. 4. Curves of the multi-cycle fatigue of steel 1045: uncoated (1), after surface modification with a memory alloy of the form TiNiZr (2) - a); The dependence of the wear rate of the layer I of TiNiZr on the pressure of the disk P at a disk sliding speed of 0.5 m / s -1; 1 m / s -2; 1.5 m / s -3; 2 m / s -4 - b).

The increase in the durability of samples with a nanostructured TiNiZr surface layer is explained both by the features of nanostructured materials destruction consisting in inhibition of fracture at the grain boundaries, preventing branching and movement of cracks due to hardening of the boundaries, as well as by the pseudoelasticity of the TiNiZr surface layer inherent to materials with thermoelastic phase transformations. The deformation accumulated by the surface layer is able to recover during cyclic unloading and, therefore, does not accumulate in the material, which increases its durability.

3 Summary

This paper developed an energy criterion for estimating the longevity and damaging of materials, which have been surface-modified by alloys with thermoelastic phase transformations, which takes into account: the energy consumed for the formation of the failure micro-surface in one cycle, in the light of the shape of the hysteresis loop; energy intensity of the alloy, in terms of phase composition according to the results of X-ray diffraction analysis; structural factor of the alloy, which is the coefficient of thermomechanical activity, determined on the basis of fractal-synergetic approach based on the results of multifractal parameterization of the structure and taking into account the influence of fractal dimension, homogeneity, periodicity and structure adaptability; calculations and diagrams of the energy capacity and theoretical strength of the TiNiZr ternary systems based on thermodynamic characteristics and phase diagrams of ternary systems.

After high-velocity oxygen-fuel spraying in the protective atmosphere of a mechanically activated TiNiZr-based shape memory powder, the cyclic durability in conditions of multicyclic fatigue increases by ~ 36%, and wear resistance is 3.5-3.6 times compared to that of steel 1045.

This work was supported by a grant from the Russian Science Foundation No. 15-19-00202 and a grant from the President MK-5445.2016.8.

References

1. K. Tanaka, S. Nagaki, Ingenieur-Archiv, **51**, 13, (1982)
2. K. Tanaka, Res. Mech, **18**, (1986)
3. C. Liang, C.J. Rogers, Eng. Math, **26**, (1992)
4. D.C. Lagoudas, Z. Bo, M.A. Qidwai, Mech. Composite Mater. Struct., **3**, (1996)
5. Zh.M. Blednova, Factory laboratory, **7**, (1988)
6. Zh.M. Blednova, The Author's abstract of the thesis of a Dr.Sci.Tech. (1989)

MATEC Web of Conferences **142**, 03006 (2018) https://doi.org/10.1051/matecconf/201814203006
ICMAE2017

Microstructure and Phase Composition of Titanium Coatings Plasma Sprayed with a Shroud

Hong Zhou[1*], *Zhi* Liu[2] and *Liancong* Luo[3]

[1] Centre for Engineering, Waikato Institute of Technology, Hamilton 3240, New Zealand
[2] Waikato Regional Council, Hamilton 3240, New Zealand
[3] Nanjing Institute of Geography and Limnology, Nanjing 210008, China

Abstract. Excellent corrosion resistance has made titanium a preferred material for use in industry with harsh environments. Protective titanium coatings combined with cheap bulk materials are expected to enable sufficient corrosion protection and lower the cost. Plasma spray processing of titanium feedstock becomes a challenging topic of research as titanium powder particles have to experience high temperatures in the flight during the thermal spraying process in an open air environment. In this paper, plasma spraying of titanium has been carried out by using a shroud in order to shield molten particles from reacting with surrounding environment. Microstructure analyses on the titanium coatings were performed by using SEM, EDS and XRD. The results show that the shroud attachment is capable of withstanding high plasma temperatures; the as-sprayed shrouded titanium coating exhibits a similar phase composition as the pure titanium powder, and an enhanced microstructure with a lower porosity for the titanium coating plasma-sprayed with a shrouding system is observed.

1 Introduction

Titanium and its alloys are well recognized for their corrosion resistance and the high strength-to-density ratio [1-3]. Excellent corrosion resistance has made titanium a preferred material for use in industry with harsh environments because they offer a high chemical resistance against various corrosive media due to a dense self-sealing oxide layer formed immediately when exposed to an oxygen containing atmosphere [4-5].

Since corrosion attack is mainly limited to the outer region of a component, a promising approach to optimize both corrosion resistance and mechanical properties is the use of coating technologies. Protective titanium coatings combined with cheap bulk materials which can provide the strength and ductility are expected to enable sufficient corrosion protection and lower the cost. Thermal spraying is a generic coating technique whereby heat sources like electric arc discharge, combustion, and plasma are used to melt the powder feedstock or wire consumable, and then projected at speed onto a substrate to build up a coating [6-7], as shown in Fig. 1.

* Corresponding author: zhouhong.nz@gmail.com

Production steps for titanium and its alloys have to be carried out carefully in an inert atmosphere condition for their affinity of oxygen and other gases. Plasma spray is considered to be the most flexible coating technique. However, air plasma spraying (APS) of corrosion-resistant coatings of titanium is difficult because an oxidation of the metal increases coating porosity and degrades coating properties. In order to prevent titanium from too much oxidation the spray process has to be carried out in vacuum or inert atmosphere. Within the thermal spray area, low pressure plasma spray (LPPS) or vacuum plasma spray (VPS) has been the promising technique used for titanium deposition thanks to its inert atmosphere and ability to form dense coatings with low porosity and oxygen content [8]. Corrosion resistance similar to that of bulk Ti has been reported if proper spraying parameters including the size of the feedstock powder are optimized [9-10]. However, this process has limitation of the size of components and is rather expensive.

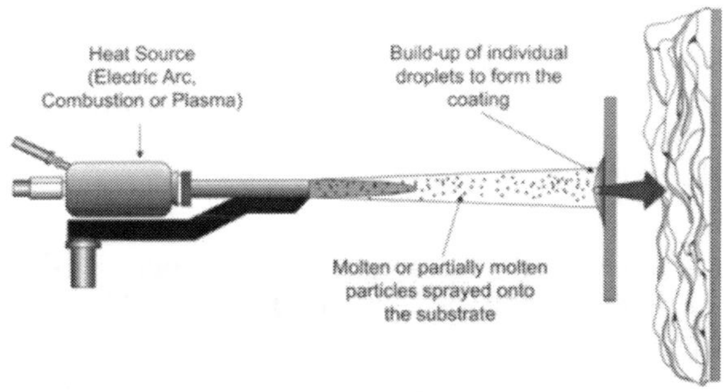

Fig. 1. Schematic diagram of coating formation process during thermal spray.

Shrouded plasma spray can be considered as a useful technology to produce low oxygen containing titanium coatings. This is to modify plasma spray torches by using an attachment onto an atmospheric plasma spraying torch [11-14]. In this work, plasma spraying of titanium has been carried out by using a shroud in order to shield molten particles from reacting with surrounding environment. Microstructure analyses on the plasma sprayed titanium coatings with/without a shrouding system were performed by using Scanning electron microscope (SEM), energy dispersive X-ray spectroscopy (EDS) and X-ray diffractometer (XRD).

2 Experimental procedures

2.1 Shroud design

A new design of shroud is proposed for plasma spray of titanium coatings, as presented in Fig. 2. The performance of a conventional conical shroud is not satisfactory due to the entrainment of air at the exit region [11]. The new design is based on a modification of the conventional shape by optimizing the profile of nozzle to a conical shape with a straight rear end. Transverse gas holes evenly distributed around the exit region help prevent air entrainment, and a large chamber for plasma jet is permitted to reduce a build-up of melted powders onto the inside wall of the shroud. A continuous circular slot at the end of shroud can make a continuous gas curtain around the plasma jet, and then further prevent the air entrainment. The new shroud was tested and plasma sprayed titanium coatings by using this shroud were fabricated.

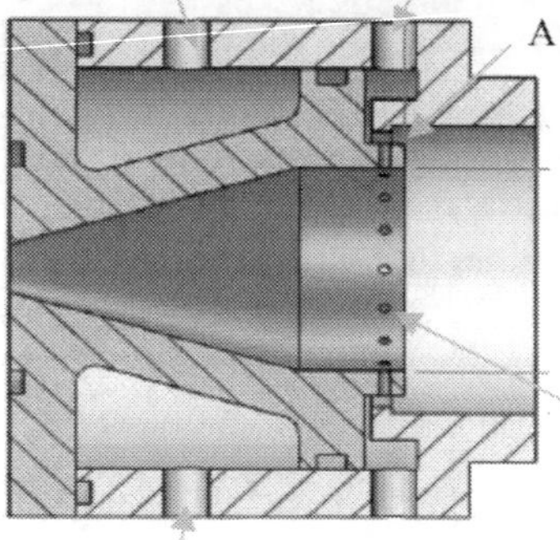

Fig. 2. Schematic cross section illustrating the key features of the shroud.

2.2 Materials

Commercially available pure titanium powder (Xi'an Lilin International Trade Co., Ltd., Xi'an, China), with particle sizes range from 45 μm to 75 μm was used as feedstock powder for the titanium coating deposition. The titanium powder is made by hydride-dehydride (HDH) method. The chemical composition of HDH Ti powder is shown in table 1. Mild steel plates (100 mm×25 mm×3 mm) were used as the substrates which were degreased and sand blasted before depositing titanium coatings onto them.

Table 1. Chemical composition of HDH titanium powder.

Element	H	O	N	C	Fe	Ti
HDH Ti powder	0.23	0.35	<0.03	0.07	<0.11	Bal

2.3 Methods and microstructural characterization

The titanium powder was then thermal sprayed by using a SG-100 plasma gun (Praxair surface technologies, USA). The novel shroud attachment was used for the shrouded plasma spraying process. The parameters for spraying with and without the shroud attachment were the same as indicated in table 2, except that there is a shroud gas in the shrouded plasma spraying process. The trial for spraying with the shroud used argon as a shrouding inert gas at the exit with a flow rate of 300 slpm.

Table 2. Spraying parameters for titanium coatings with and without a shroud.

Spray parameter	*setting*
Current, A	800
Voltage, V	80
Primary gas, Argon, slpm	85
Auxiliary gas, Helium, slpm	18
Powder feed rate, g/min	30
Spray passes	10
Spray distance, mm	100
Transverse speed, mm/s	500
Shroud gas, Argon, slpm	300

Cross sectional samples were metallographically prepared for microstructural analysis. Scanning electron microscope (SEM, Hitachi S4700, Japan) was used to observe the microstructures and morphology of specimens. An energy dispersive X-ray spectroscopy (EDS, Oxford INCA, UK) was used to detect the chemical compositions on a microscopic scale. Phase transformations were studied using an X-ray diffractometer (XRD, Bruker, model D8, Germany) with Cu*K*α radiation at 40 kV and 30 mA, the scanning speed of 2θ was 2° min^{-1} when measuring from 20° - 80°.

3 Results and discussions

3.1 Phase composition

The phase composition result of the titanium feedstock powder is shown in Fig. 3a. From the XRD pattern, it can be found that the HDH titanium powder is merely composed of α-Ti phase. Fig. 3b and 3c present the XRD patterns for titanium coatings with/without the shroud system. For the unshrouded titanium coating, it is composed of a mixture of phases including α-Ti, TiO and TiN. The result indicates that oxidation and nitrodization occurred on the pure titanium particles at high temperatures during the air plasma spraying process. As there is no shroud protection, the oxide or/and nitride might be formed when the oxygen and nitrogen from surrounding air came in contact and reacted with molten titanium particles in flight. From the point of melting in the plasma until impact on the substrate, each molten or partially molten particle is prone to oxygen/nitrogen absorption/dissolution and oxidizing gas–liquid metal oxidation reactions. Oxygen/nitrogen from the entrained air is dissociated into highly reactive monotonic oxygen/nitrogen at high temperatures. This enable titanium particles oxidation/nitrodization to occur in a very short time. Oxidation may also take place during splat formation and coating build-up, even after solidification of the splats since the surface of each hot splat is prone to oxidation by the high temperature.

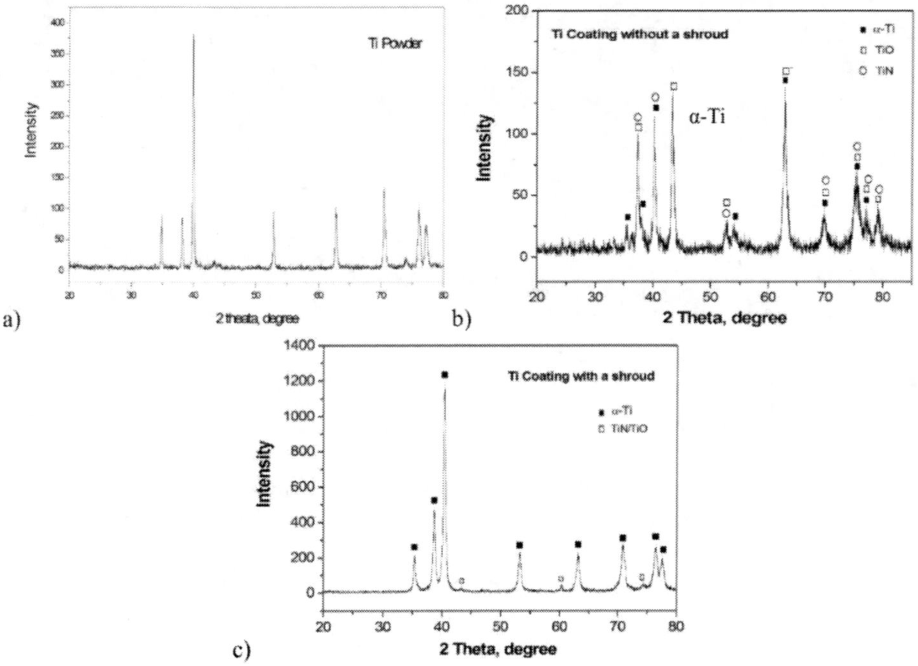

Fig. 3. XRD patterns of the titanium powder (a) and titanium coatings plasma-sprayed in air without (b)/ with (c) the shroud.

However, for the titanium coating plasma-sprayed with a shrouding system, the XRD pattern presents a similar curve as that for the pure titanium powder. The result reveals that the dominant phase in the shrouded titanium coating plasma-sprayed is still composed of α -Ti, which means that lower oxygen content in the shrouded titanium coating is obtained. Meanwhile, some very weak peaks for oxide and nitride of titanium can be found from the pattern, which means that there might exist a trace of oxide and nitride in the shrouded titanium coatings.

3.2 Microstructure

Fig. 4 presents the second electron images of polished cross sections of the titanium coatings with/without the shroud. The SEM images reveal that the shrouded titanium coating has a dense microstructure with a very low porosity; whereas the air plasma sprayed titanium coating possesses a high porosity. The presence of the shroud leads to a significant reduction in coating porosity. In principle the reduction in air entrainment with the shroud was expected to result in a better heating of the particles and then more molten titanium particles at higher temperatures in flight were obtained, which help decrease the porosity in the titanium coatings. As we know, the attachment itself shielded the plasma jet as it exited the spray gun in the shrouded plasma spraying process, and then a gas shroud enveloped the plasma jet as it left the attachment. The additional inert shroud gas was injected around and enveloped the plasma jet, shielding the molten particles from reacting with the surrounding environment, reducing the amount of entrainment of cold and heavy ambient air and delaying the corresponding drop in velocity and temperature, and thus retarded particle cooling and reduced oxygen content and porosity of the resulting coatings. Meanwhile, gas-shrouded nozzles could extend the hot core of the jet by rearranging the gas flow to increase the dwell time of particles in the plasma.

a) b)
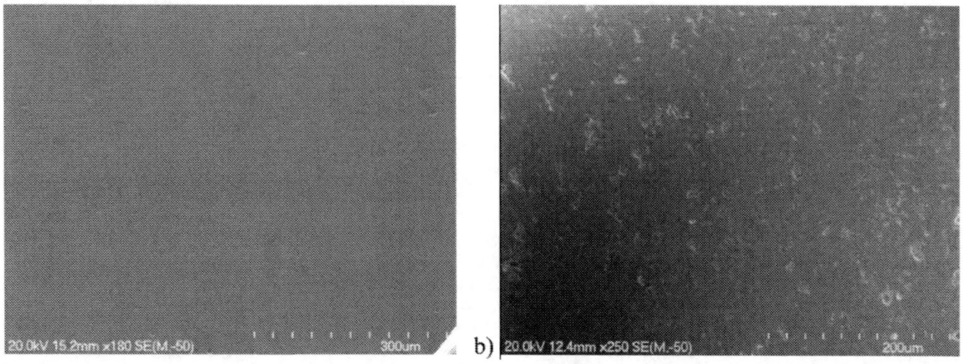

Fig. 4. SEM images for plasma sprayed Ti coatings with (a) and without (b) the shrouded.

Fig. 5 shows the SEM morphologies and element area compositions of the air/shrouded plasma sprayed titanium coatings, which was determined by EDS. The Ti and O contents of air/shrouded plasma sprayed titanium coatings are given in table 3. The EDS analyses indicate that the shrouded titanium coating presents an ideal pure titanium, and oxide and nitride were not found in the area assessed by EDS. However, oxygen element with 18.04±1.28 weight% is determined in the air plasma sprayed titanium coating, titanium is 81.12±1.28 weight %, and other elements take the balance. This result also verifies that oxidation occurred in the air plasma spraying process; and the shrouding system provides an effective shielding to prevent oxidation from surrounding air entrainment.

Table 3. Ti and O contents of titanium coatings determined by EDS.

	Shrouded		*Unshrouded*	
Element	*Ti*	*O*	*Ti*	*O*
Weight%	100.00	*0*	*81.12*	*18.04*
Weight % Error (+/- 1 Sigma)	+/- 0.75	-	+/- 0.75	+/- 1.28

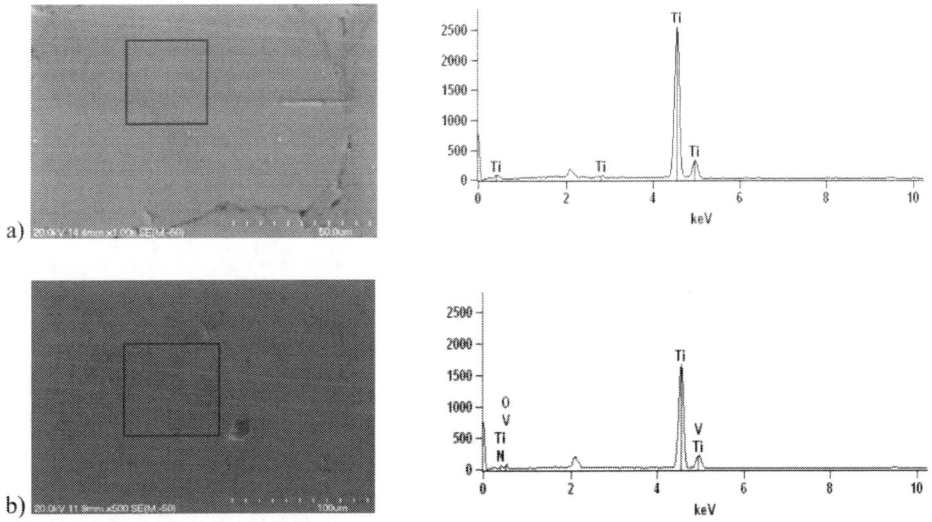

Fig. 5. SEM morphology and element area compositions of the air (a)/shrouded (b) plasma sprayed titanium coatings.

4 Conclusions

Titanium and titanium alloys have been used more and more for corrosion protection. Plasma spraying with a shrouded is a useful technology to produce low oxygen containing titanium coatings. The results show that the shroud attachment is capable of withstanding high plasma temperatures; oxidation and nitrodization occur on the pure titanium particles at high temperatures when the powder feedstock is air plasma-sprayed; the as-sprayed shrouded titanium coating exhibits a similar phase composition as the pure titanium powder, which means that a lower oxygen content in the shrouded titanium coating is obtained; and an enhanced microstructure with a lower porosity for the titanium coating plasma-sprayed with a shrouding system is observed, the shroud lead to a significant reduction in coating porosity.

References

1. C.Leyens, M.Peters, *Titanium and titanium alloy.*,16-18(Wiley-VCH, Weinheim,(2003)
2. R. Boyer. Mater. Sci. Eng., **A213**, 103-114 (1996)
3. H. Zhou, B. Gabbitas, S. Matthews, D. Zhang, *Proceedings of the 12th World Conference on Titanium*, **V3,** 1906-1910 (2011)
4. K. Luthra, Oxidation of Metals, **36,** 274-290 (1991)
5. S. Taniguchi, T. Shibata, S. Itoh, Materials transaction, **32,** 151-156 (1991)
6. P. Fauchais, G. Montavon, G. Bertrand, J. Thermal Spray Technol., **19,** 56-80 (2010)

7. P. Fauchais, M. Fukumoto, A. Vardelle, J. Thermal Spray Technol., **13**, 337-360 (2004)
8. J. Kawakita, S. Kuroda, T. Fukushima, H. Katanoda, Surf. Coat Technol., **201**, 1250–1255 (2006)
9. H. Steffens, E. Erturk, K. Busse, J. Vac. Sci. Technol., **A3**, 2459-2463 (1985)
10. E. Lugscheider, P. Lu, B. Haeuser, D. Jaeger, Surf. Coat. Technol., **32**, 215-226 (1987)
11. M. Jankovic, J. Mostaghimi, Plasma Chem. Plasma Process., **15**, 607–628 (1995)
12. M. Morks, C.C. Berndt. Applied Surf. Sci., **256**, 4322-4327 (2010)
13. M. Planche, H. Liao, C. Coddet, Surf. Coat Technol., **202**, 69–76 (2007)
14. M. Jankovic, J. Mostaghimi, V. Pershin, J. Thermal Spray Technol., **9**, 114-120 (2000)

Anti-carburizing Coating for Resin Sand Casting of Low Carbon Steel Based on Composite Silicate Powder Containing Zirconium

Chunyi Zhan [1,2], *Shengshan* Feng [1,2*], *Shuzhong* Xie [3,4], *Chunjing* Liu [3,4], *Yunhua* Gao [3,4], and *Jiahao* Liang [3,4]

[1]College of Mechanical and Electrical Engineering, Guangdong Polytechnic of Science and Technology, Zhuhai 519090, China
[2]Zhuhai Key Laboratory of Advanced Equipment Manufacturing and Material Processing Technology, Zhuhai 519090, China
[3]Zhuhai Doumen Fulian Molding Materials Industry Co., Ltd, Zhuhai 519080, China
[4]Guangdong Molding Material Engineering Technology Center, Zhuhai 519080, China

Abstract. This paper studied the structure and properties of anti-carburizing coating based on composite silicate powder containing zirconium by X-ray diffraction analyzer, thermal expansion tester, digital microscope and other equipment. It is introduced that the application example of the coating in the resin-sand casting of ZG1Cr18Ni9Ti stainless steel impeller. The anti-carburizing effect of the coating on the surface layer of the cast is studied by using direct reading spectrometer and spectrum analyzer. The change of the micro-structure of the coating after casting and cooling is observed by scanning electron microscope. The analysis of anti-carburizing mechanism of the coating is presented. The results indicate that the coating possesses excellent suspension property, brush ability, permeability, levelling property and crack-resistance. The coating exhibits high strength and low gas evolution. Most of the coating could be automatically stripped off flakily when the casting was shaken out. The casting possesses excellent surface finish and anti-metal penetration effect. The carburizing layer thickness of the stainless steel impeller casting with respect to allowable upper limit of carbon content is about 1mm and maximum carburizing rate is 23.6%. The anti-carburizing effect of casting surface is greatly improved than that of zircon powder coating whose maximum carburizing rate is 67.9% and the carburizing layer thickness with respect to allowable upper limit of carbon content is greater than 2mm. The composite silicate powder containing zirconium coating substantially reduces the zircon powder which is expensive and radioactive and mainly dependent on imports. The coating can be used instead of pure zircon powder coating to effectively prevent metal-penetration and carburizing of resin-sand-casting surface of low carbon steel, significantly improve the foundry production environment and reduce the production costs.

* Corresponding author: fss8172@126.com

1 Introduction

Resin sand process is one of the main methods for producing carbon steel castings. However, it is easy to cause carburization defects in the surface of low carbon steels. The pearlite increases, ferrite decreases and acicular martensite appear in its microstructure, which leads to the decrease of plasticity, toughness, machining and welding performance [1-5]. To solve these problems, it is necessary to form a coating with a certain thickness of dense sintering layer on the surface of resin sand mold (core) under high temperature. Meanwhile, adhering sand and carburizing of casting surface are avoided.

Zircon powder is now widely used as a refractory material for anti-carburizing and anti-sand burning coating for low-carbon steel cast at home and abroad [4-6]. But its mineral is scarce and expensive. Moreover, monazite contains low concentrations of uranium and thorium oxides, it has a certain intensity of radioactivity and can produce ionizing radiation, which can cause excitation and ionization of biological macromolecules, cause DNA damage and increase chromosome aberration rate [6, 7]. Therefore, it is imminent to study and search for substitute materials of zircon powder.

In this paper, low-cost composite silicate powder refractory material containing zirconium is used to develop new coatings for steel castings, which has weak or no radioactivity, lower cost of production, which can be automatically peeled off, and has a good effect on both on anti-carburizing and anti-metal-penetration. It can meet the urgent needs of low carbon steel casting, which has low production cost, high quality and environmental friendliness.

2 Composition design of composite silicate powder anti-carburizing coating containing zirconium

Foundry coating is commonly composed of refractory aggregate, binder, sintering agent, suspending agent, solvent, assistant and other components.

2.1 Refractory aggregate

Refractory aggregate is the skeleton of foundry coating. It should have high temperature and heat resistance in order to avoid the formation of mechanical sand castings, and also should have stable chemical property at high temperature in order to avoid interfacial chemical reaction with sand mold or core and molten metal.

In this experiment, composite silicate powder containing zirconium was selected as the main refractory aggregate. It is an aluminum silicate based composite silicate with milky white color, whose chemical is as follows: $Al_2O_3 \geqslant 50\%$,$SiO_2 \leqslant 35\%$,$Fe_2O_3 \leqslant 1\%$,$TiO_2 \leqslant 5\%$,$ZrO_2 \leqslant 9\%$.

According to the test results of specific activity of radionuclide of composite silicate powder containing zirconium show that, the specific activity of U is 1.671Bq/g and Th is 0.413Bq/g, about half the specific activity of radionuclide in Australia and South Africa (U 2.0Bq/g ~4.0Bq/g, Th 0.1Bq/g ~1.0Bq/g). It has good safety and environment-friendly.

The X ray diffraction petrographic analysis map of the composite silicate powder containing zirconium is shown in Fig. 1. XRD analysis results show that the lithofacies of the composite silicate powder containing zirconium is mainly corundum, quartz, zirconite and kyanite.

The coefficient of linear expansion of the kyanite in the composite silicate powder containing zirconium in the heating process is high. When heated to about 1100 degrees celsius, it begins to decompose into mullite and ristobalite (glass polymer) and volume dilation occurs at the same time. When heated to over 1300 degrees celsius, it decomposes rapidly. When heated to over 1450 degrees celsius, it decomposes completely in the main and totally transform into mullite and ristobalite.

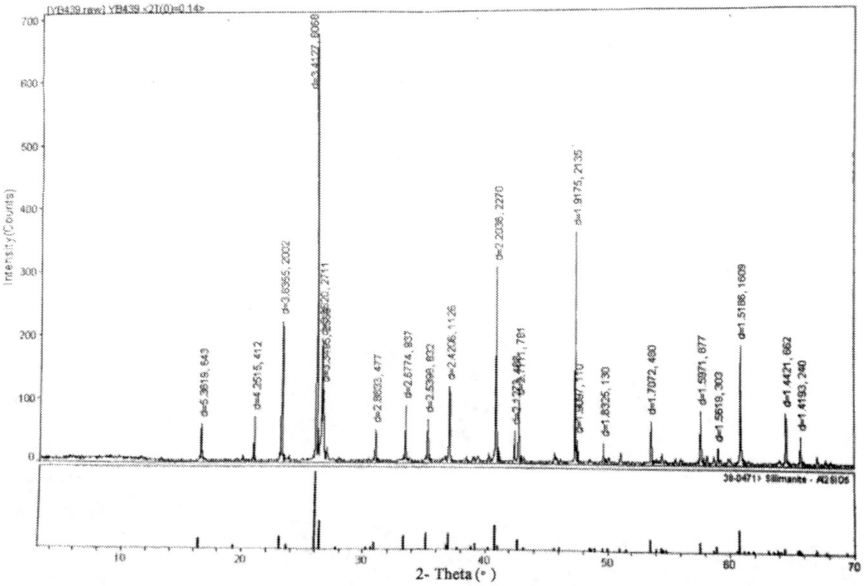

Fig. 1. The X ray diffraction petrographic analysis map of the composite silicate powder containing zirconium

Mullite has high refractoriness (stable at 1800℃) and chemical inertness. The transformation process of from kyanite to mullite at high temperatures is irreversible, which is stable below 1810℃. And when temperature decreases, the volume is stable and the shrinkage is low.

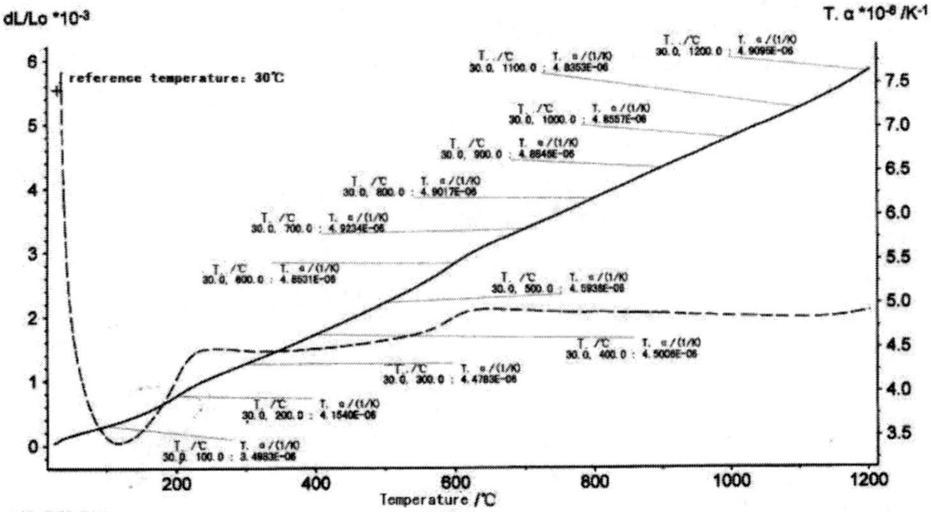

Fig. 2. The thermal expansion curve of the composite silicate powder containing zirconium

Fig. 2 shows the thermal expansion curve of the composite silicate powder containing zirconium. Kyanite increases the difference of the linear contraction coefficient between the composite silicate powder containing zirconium coating and casting metal, which produce greater shear stress between casting and coating boundary and can lead to self-stripping in layers between coating and casting. The cristobalite with high viscosity (glass polymer) in the process of transformation process of kyanite at high temperatures can resist the permeation of molten metal. It is helpful to prevent the mechanical burn-on defect of steel castings. And since the difference of the linear contraction coefficient between the glass coating layer and the casting metal is large after cooling, which increase shear stress between casting and coating boundary and further lead to self-stripping in layers between coating and casting, thereby obtaining castings with smooth surface. The thermochemistry stability of generated mullite is strong. The mullite is not easy to react with acid and alkaline oxides that are helpful to prevent the chemical burn-on defect of steel castings.

The price of the composite silicate powder containing zirconium is only 1/3~1/4 that of zircon powder. But particle shape of the composite silicate powder containing zirconium is mainly polygonal (see Fig. 3), which is a little worse than that of spherical zircon powder (see Fig. 4).

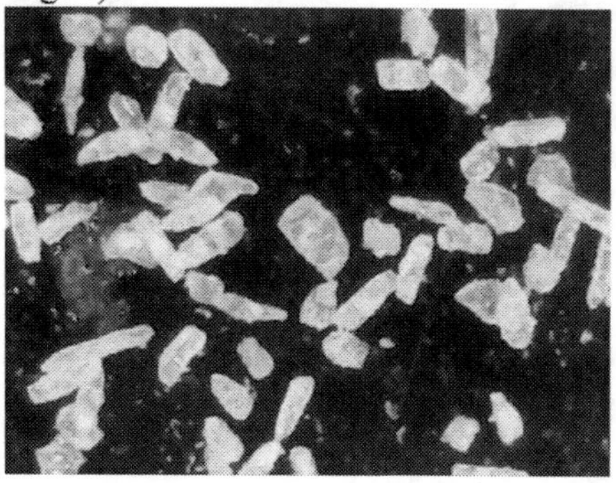

Fig. 3. The particle morphology of the silicate powder containing zirconium(200╳)

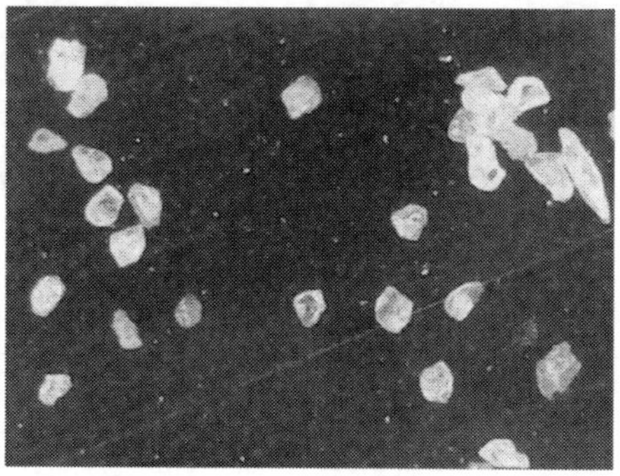

Fig. 4. The particle morphology of zircon composite powder(200╳)

2.2 Sintering agent

The sintering agent should ensure that the sinter can form a continuous dense sinter skin, which has a significant ability to prevent burn-on and shield carbon. In this study, fluoride was selected as sintering agent.

The sintering mechanism of fluoride is that the electrostatic potential of F- introduced by fluoride is only about 1/2 that of O2-, and the radius of both ions are very close. F- reacts with aluminum-oxygen or silicon-oxygen network in refractory aggregate, which can break the network. The equation is as follows:

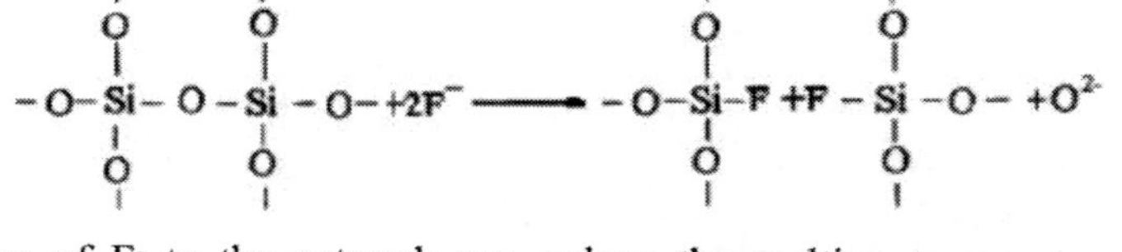

The damage of F^- to the network can reduce the melting temperature of refractory aggregate, but excessive fluoride can reduce its damage to the network. The experimental results show that the suitable dosage of fluoride is 1%~3% amount of the refractory aggregate.

In addition, a sintering aid is adopted. It not only can promote the sintering of coating layers, but also can decompose at temperatures over 1000℃, which can produce a reducing gas with good stability at elevated temperature and high vapor pressure. This reductive gas and sintered compact coating together can avoid contact of solid carbonaceous material in the sand mold (core) / casting interface with hot metal, and can prevent solid diffusion of carbon to layer of molten steel interface, thus effectively reduces or eliminate the surface carburization of castings.

2.3 Particle size and gradation

The smaller the radius of such powder material as the refractory aggregate and the sintering agent etc., the better the suspension property and brush ability, the greater the penetration depth of the coatings to the sand mold, the better the shielding effect of the coating on the carbon, the lower the adding amount of suspending agent needed. The finer the particle size of sintering agent, the easier the sintering of the coating, the higher the density, the better the shielding effect on carbon. But if the particle size of powder is too small, it will make the coating easily be cracked, and affect anti-metal-penetration and anti-carburizing ability. Therefore, the particle size of such powder material as the refractory aggregate and the sintering agent etc. should be moderate, which is appropriate between -200 and 500 meshes.

Refractory aggregate must also have a reasonable particle gradation, and the size distribution should show large dispersion within the appropriate particle size range to make full use of mutual mosaic effect of coarse and fine particles, reduce the inter particle porosity in order to obtain compact coatings, improve its anti-metal penetration and anti-carbonization ability.

2.4 Binder

Foundry coating not only has a certain normal temperature strength, but also have good high temperature strength, so the binder should adopt a composite process using organic binder at low temperature together with inorganic binder at high temperature. However, the organic binder generally contains carbon, and its pyrolysis products are bound to contain carbon. Therefore, in order to ensure that the coating have a certain normal temperature strength, the adding amount of organic binder in coating should be minimized and a little more inorganic binder at high temperature should be used. The coating use normal temperature air-dried organic binder at low temperature together with inorganic binder at high temperature as binder [8].

2.5 Solvent

There are two kinds of commonly used solvents: water and alcohol. When alcohol is used as solvent, binder is generally organic binder. Since pyrolytic carbon produced by alcohol solvent and organic binder in the casting process remain in the coating, which will increase the tendency of carburizing of casting. Therefore, the coating uses water as solvent.

3 Main properties of composite silicate powder anti-carburizing coating containing zirconium

Tested by Guangdong province casting molding material engineering technology center, the main technical properties of new composite silicate power anti-carburizing material containing zirconium (water-based and air-dry) is shown in table 1. Brushing index in the table refers to the ratio of apparent viscosity η 6 at low shear rate (6r/min) and η 60 at high shear rate (60r/min), which characterizes shear thinning or thickening property of coatings. The higher the index, the better the brushing property of coating, which means easy operation of brushing, good anti-flowing property and easy thick coating.

Thixotropy rate refers to the change rate of the apparent viscosity of coating η with the change of stirring time. The higher the thixotropy rate, the better the leveling property, the greater the dilution range, the better the brushing property, the worse the anti-flowing property.

Table 1. Technical performance of composite silicate powder anti-carburizing coating containing zirconium.

Solvent	Conditional viscosity s	Density $g \cdot cm^{-3}$	Suspension property(24h) %	Thixotropy rate %	Brushing index	Gas evolution $ml \cdot g^{-1}$	Hot cracking grade
water	12	1.92	99	37.2	7.6	12.7	1

4 Application effects of coatings on stainless steel castings

The trial casting is stainless steel impeller, whose material grade is ZG1Cr18Ni9Ti. The casting weight is about 30Kg, the wall thickness is 15mm ~ 22mm and the tapping temperature of molten steel is about 1700℃. Acid-cured furan resin self-hardening sand was used to make cores, at the same time, two identical rectangular specimen sand mold were manufactured, whose size is 50×90×110mm. Two kinds of casting sand mold and block sand were used to compare the water based self-drying coatings containing zirconium and silicate powder and coatings. Water-based and self-drying composite silicate powder coating containing zirconium and water-based and self-drying zircon powder coating were used separately in two casting sand mold and specimen sand mold for contrast. The total thickness of the coating is 0.5 mm ~0.8mm and it doesn't need to be dried before casting, so as to avoid the influence of residual carbon produced by flame combustion on coating surface on surface carburization of castings.

A line cutting machine was used to intercept section specimen at the same position of the specimen, and chemical composition of the center and surface of direct reading spectrometer test block were used separately.

The experiment results show that the suspension property of the new anti-carburizing composite silicate powder alcohol based coating containing zirconium is good, is easy to brush and is good in levelling property. The coating of castings and blocks produced using the coating was basically automatically stripped from the casting surface flakily; no defects such as sand burning, porosity, sand inclusion, coating inclusion etc. were seen on the casting surface.

Table 2 shows the spectral analysis results of chemical composition of two kinds of specimen (comparison of two coatings) at different distances below the surface.

The original carbon content of molten steel is 0.106%~0.110%, and the carbon content of the surface layer of castings should be controlled within 0.12% after pouring.

It can be known from table 2 when using the new composite silicate powder anti-carburizing coating containing zirconium that the carbon content of the surface of the test block is 0.136% and maximum carburizing rate is 23.6%. The carbon content of the upper limit of 0.12% of the carburizing layer thickness is about 1mm, the anti-carburizing effect is significant. And when using a domestic zircon powder composite coating, the carbon content of the surface of the test block is up to 0.178%, exceeding the allowable upper bound of carbon content 0.12%, which does not reach the quality requirements of manufacturing. The highest carburizing rate is 67.9%, about three times more than the new coating. The thickness of the carburized layer corresponding to the allowable upper limit of carbon content is more than 2mm.

Table 2. Spectrum analysis results of chemical composition at different depth under the surface of stainless steel impeller casting specimen

Types of coatings	position	C / %
composite silicate powder coating containing zirconium	Specimen surface	0.136
	Under specimen surface 1mm	0.122
	Under specimen surface 2mm	0.115
	Center of specimen	0.110
zircon powder coating	Specimen surface	0.178
	Under specimen surface 1mm	0.137
	Under specimen surface 2mm	0.132
	Center of specimen	0.106

Fig. 5 is the result of energy spectrum analysis of carbon content increment ratio, which is beneath the surface layer within 6mm depth and two kinds of specimen, whose left and right sides are using a domestic zircon powder composite coating and the new composite silicate powder anti-carburizing coating containing zirconium, which is cast by the molten steel of the same furnace and is measured by Shimadzu Corporation EPMA-1720 energy spectrum analyser. The results show that it basically agrees with the results of using the spectrometer test, this further proved that the new coating has a good effect of anti-carburizing.

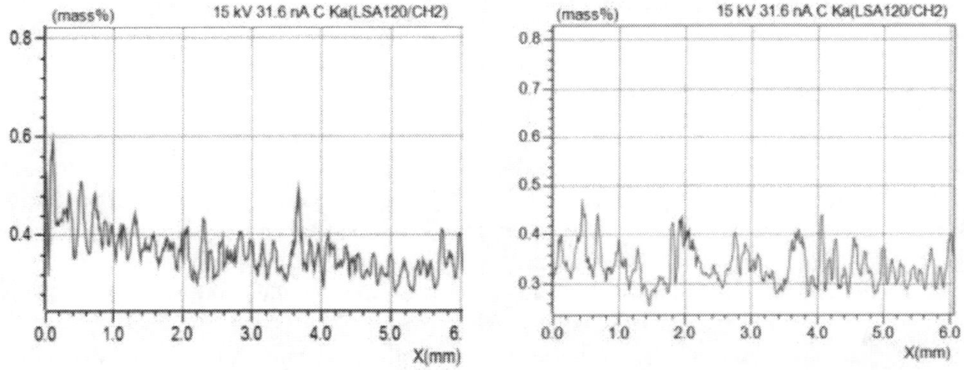

Fig. 5. The results of the spectrum analysis of the specimen surface (incremental ratio of carbon content (Left: Using zircon powder coating Right: Using composite silicate powder coating containing zirconium)

5 The anti-carburizing and anti-metal penetration mechanism of composite silicate powder coating containing zirconium

Scanning electron microscope (SEM) was used to observe the changes of the microstructure of the coating after casting and cooling. The results are shown in Fig. 6.

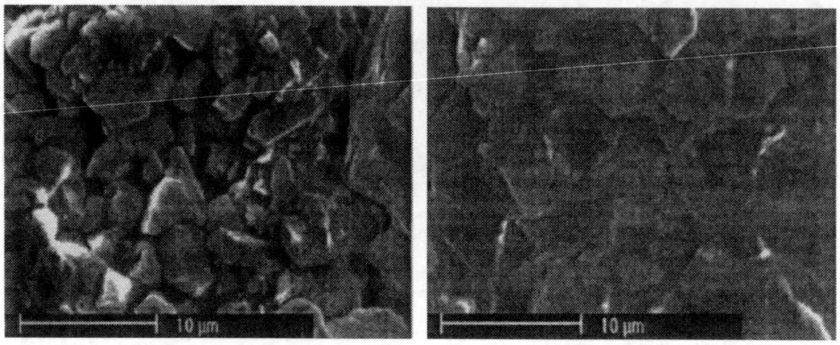

Fig.6. The SEM morphology of sinterskin surface of composite silicate powder coating containing zirconium (Left: Using zircon powder coating Right: Using composite silicate powder coating containing zirconium)

It can be seen from Fig. 6 that an irreversible phase change and great expansion occurs in the kyanite in the zircon powder coating at lower temperature compared with those in zircon powder coating. It makes the powder particle surface in the sinter-skin of coating sintered more fully. Common adhesion between particles occurs due to surface melting, the coating exist as glassy phase. The gap of the coating is small and the density of coating is high, which has a strong shielding effect on the migration and diffusion of elements. This is the reason why the silicate powder coating containing zirconium has good anti-carburizing and anti-metal penetration performance.

6 Conclusions

(1) The permeability, anti-flowing property and crack resistance of low carbon steel resin sand casting anti carburizing coating based on composite silicate powder containing zirconium are very good. The coating exhibits high strength and low gas evolution. Most of the coating can be stripped out flakily when the casting is shaken out, the casting surface is clean and the effect of anti-metal penetration is good. There are no defects such as porosity, sand inclusion, coating inclusion etc. The cleaning efficiency of casting is greatly improved, energy is saved and labor intensity is decreased.

(2) The thickness of the carburizing layer of composite silicate powder coating containing zirconium relative to the allowable limit of carbon content is about 1mm, whose maximum carburizing rate is 23.6%. Anti-carburizing effect of the casting is significantly higher than that of zircon powder coating, whose maximum carburizing rate is 67.9% and the thickness of the carburizing layer relative to the allowable limit of carbon content is more than 2mm, which can basically meet the production demands of low-carbon steel castings such as stainless steel.

(3) The zirconium containing composite silicate powder coating can greatly reduce the cost of the coating and the cost of casting production, because it can greatly reduce the cost of the zirconium powder, which is of high cost and has a certain radioactivity. It also can improve foundry production environment and help to protect the health of workers.

In this paper, the research was sponsored by Industry-university-research Cooperation Projects of Zhuhai City, Guangdong Province, China (Project No. 2012D0501990013).

References

1. L. Li, R. Tan, H.K. Li, et al, Study on the Anti-Carburizing Coating Applied to Furan Resin Sand, Foundry. **63**, 386(2014)

2. L. Zhao, Y. Tian, Y.X. Wang, et al, Effect of Furan Resin Sand on Carburization Defects of Ultra-Low Carbon Steel Casting, Foundry. **63** , 929(2014)
3. T. Z. Li, Effect of Furan Resin Bonded Sand Mold on Surface Microstructure of Steel Castings, Foundry. **59**, 1069(2010)
4. S.S. Feng, Investigation and Application of Carbon-impervious Coating on the Surface of Low-cabon Steel Castings, Foundry Technology. **31**, 1551(2010)
5. Z. Zhang, M. Chen, Application of Shielding Coating to Steel Casting Process, Foundry. **59**, 940(2010)
6. Z.Z. Wu, C.C. Yang, The Radioactivity of Zircon Sand and its Harmful to Human Bodies, Industrial Safety and Environmental Protection. **29**, 20(2003)
7. S.Y. Wang, B.J. Zhangn,C.T. Du, et al, Study on DNA Damage Effect of Zircon Powder on Foundry Workers, Chinese Journal of New Clinical Medicine. **4**, 563(2004)
8. C.Y. Zhan, S.S. Feng, S.Z. Xie, et al, Curing properties of water-based self-drying fast-drying foundry coating, Proceedings of the 2015 3rd international conference on material, mechanical and manufacturing engineering. **27**, 1984(2015)

Preparation and Characterization of FC Films Coated on PET Substrates by RF Magnetron Sputtering

Mei-lin Huang[1, 2] , *Sheng-guo* Lu [1*], and *Wen-qin* Du[2]

[1]Guangdong Provincial Research Center on Smart Materials and Energy Conversion Devices, Guangdong Provincial Key Laboratory of Functional Soft Condensed Matter, School of Materials and Energy, Guangdong University of Technology, Guangzhou 510006, Guangdong, China
[2]Guangdong College Research Center of Functional Textile Engineering and Technology, School of Textile and Clothing, Wuyi University, Jiangmen 529020, Guangdong, China

Abstract. Fluorocarbon (FC) films were prepared on polyethylene terephthalate (PET) plates and PET fabrics respectively by a radio-frequency (RF) magnetron sputtering technique using polytetrafluoroethylene (PTFE) as a target. Scanning electron microscope and X-ray photoelectron spectroscopy were used to investigate the morphology, structure and composition of the obtained FC films. The hydrophobicity and uvioresistant properties of the FC film coated fabric were studied. The results show that the FC films were successfully deposited on the PET substrates by a RF magnetron sputtering. The deposited films are made up of four components -CF3, -CF2-, CF- and -C-. The proportions of the four components and surface morphologies of the deposited films vary with the sputtering conditions. Compared with the original fabric samples, the hydrophobicity of the FC film coated fabrics is quite good and improved significantly.

1 Introduction

Magnetron sputtering has been taken into account as a very effective deposition method for growing thin films. It is easy to prepare high-purity and good uniformity films because of its many advantages, e.g., multiple targets, ease of sputtering parameter control, large film area, and good adhesion. The magnetron sputtering technology can be used to prepare super-hard films, wearable and corrosion-resistant films, superconducting films, magnetic films, optical films, and other specific films [1]. It has been using the magnetron sputtering to deposit thin films on fabric to obtain functionalities, such as wettability, hydrophilicity, hydrophobicity, antibacterial property, and uvioresistant property [2-9].

There have been lots of studies on the coating of fluorocarbon films by using magnetron sputtering to achieve various properties [10-15]. Nevertheless, since the properties of the deposited films are greatly affected by the sputtering conditions, e.g., sputtering power, sputtering time, vacuum pressure, substrate temperature, gas flowing rate and other factors

* Corresponding author: sglu@gdut.edu.cn

[16-18], it is of significance to investigate the relationship between the sputtering conditions and the properties of the films coated on the textiles directly.

In this work, the fluorocarbon (FC) films were deposited on the polyethylene terephthalate (PET) plates and PET fabrics, respectively, by a RF magnetron sputtering technology using polytetrafluoroethylene (PTFE) as a target. The influences of different sputtering conditions on the properties of FC films and coated fabrics were studied especially the impact of substrate's temperature. The compositions, morphologies, and microstructures of the films were analysed by a scanning electron microscope (SEM) and an X-ray photoelectron spectroscopy (XPS). The hydrophobicity of the FC film coated fabric were also tested and discussed.

2 Experimental details

2.1 Materials

The substrate material is a kind of PET fabric, which was cut into circular sample with 6 cm in diameter, and washed in acetone and ethanol successively, then vibrated for 10 min by an ultrasonic cleaner, and then repeat for several times using deionized water, and dried finally.

The PET thick plates, which have the same chemical composition as the PET fabric, with 0.35 mm in thickness, were also used as substrates, because it is easier to investigate the morphology, configuration and structure of the films deposited on the PET plates.

The target material used was commercial PTFE plates.

2.2 Deposition of FC films

The RF magnetron sputtering apparatus (Model FJL450E) with a 13.56 MHz power supply was used in the experiments. The FC films were deposited consecutively on PET plates and PET fabrics using PTFE as a target under conditions shown in Table 1. The distance between the target and the substrate was set as 60 mm, with a base pressure of 4×10-4 Pa, sputtering working pressure of 0.5 Pa, and argon (Ar) gas flowing rate of 40 ml/min.

Table 1. The sputtering conditions for deposition of FC films

Sample	Substrate	Sputtering power (W)	Substrate temperature (oC)	Sputtering time (min)
#1	PET plate/fabric	150	40	30
#2		150	60	30
#3		150	80	30
#4		210	40	120
#5		210	70	120
#6		210	100	120

2.3 Characterization

The morphologies were observed by a scanning electron microscope (SEM, JEOJ-JSM-840). The chemical compositions and structures of the deposited films were analyzed using

XPS (VG ESCA LABS5 spectrometer), in which the monochromatic Al Ka radiation was used as the X-ray source. The XPS curves were studied using a conventional analytical procedure [19,20]. The samples' hydrophobicity before and after the sputtering was tested by a static contact angle measurement (JC2000A) with conventional water droplet method and by a water repellency tester (Y(B)813) with standard GB/T4745-1997.

3 Results and discussion

3.1 Surface morphology and structure

The SEM images of the films deposited on PET plates of Sample #1 - #6 are shown in Fig. 1. Results indicate that the surface morphology of the deposited films varies with the sputtering conditions. As can be observed in Fig. 1 (a) - (c), the surfaces of Sample #1 - #3 look like mountains, with concaves and convex, grooves and buckling. The number of grains were less and the surface is not smooth because the sputtering time is shorter and the growth of film insufficient. The grain sizes was ranged from 10 nm to 1 m. There are wrinkles appeared on the surfaces of three PET plates, and the higher the substrate temperature, the worse the wrinkles are. It is obvious that the substrate temperature has less effect on the film formation but more effect on the PET plates themselves. In brief, after sputtering for 30 min in power of 150 W, the FC grains are dispersively deposited on the substrate and the FC film still has not formed.

In Fig. 1 (d) - (f) for Samples #4 - #6, results indicate that after sputtering for 120 min at 210 W, the film's thickness increases with the increased substrate temperature, which makes the surface smooth. With further deposition, contraction crack emerges and the surface becomes fragile and flaws form eventually. When the substrate temperature increases from 40 to 100°C, the surfaces of films become more compact, dense and uniform. The investigations indicate that either the sputtering time of 120 min is too long, or the sputtering power of 210 W is too high, hence the impact of substrate temperature is not obvious. Therefore, the sputtering at a relatively lower substrate temperature is better.

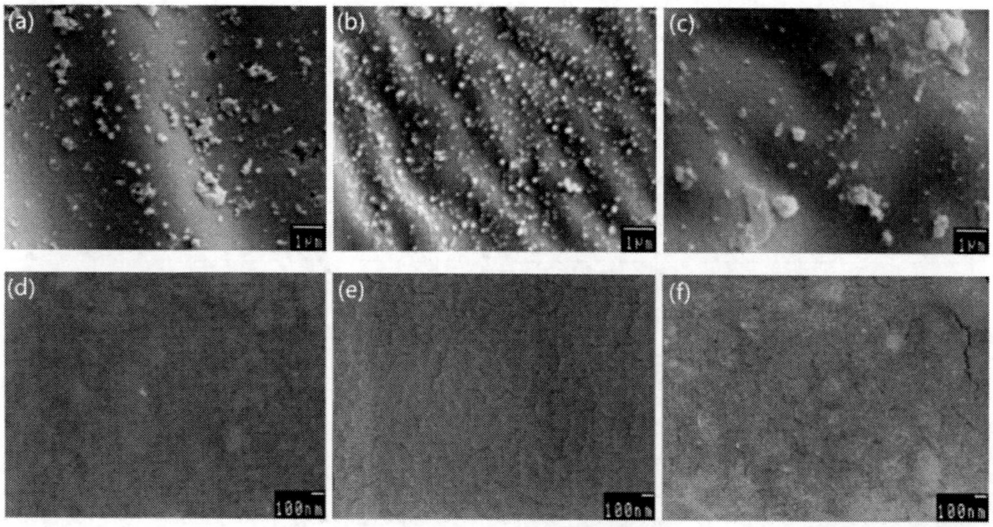

Fig. 1. SEM images of (a)150 W,40°C, 30min; (b)150W,60°C, 30min;(c) 150W,80°C, 30min; (d) 210W,40°C, 120min;(e) 210W,70°C, 120min; (f) 210W,100°C, 120min

3.2 Composition analysis and hydrophobicity study

The composition and chemical structure of the deposited films were determined from C1s spectrum which was resolved into four sub-bands -CF3, -CF2-, CF- and -C-. The software XPSPEAK was used to fit the XPS curves of C1s spectrum and obtain the position, area and FWHM (full width at half maximum) of the peaks of -CF3, -CF2-, CF- and -C-, respectively. The proportions of four components, F/C atom ratio and the static contact angle of Samples #1 - #6 are given in Table 2. A typical C1s spectrum of fluorocarbon film was shown in Fig. 2.

Table 2 Proportions of four components, F/C atom ratio and static contact angle of the FC films

Sample	Substrate	0	$-CF_2-$	-CF-	-C-	F/C ratio	Static contact angle
		(%)					(degree)
#1	PET plate	5.31	18.25	63.51	12.92	1.16	96.9
#2		19.34	5.40	25.92	49.34	0.95	94.6
#3		5.01	13.97	10.75	70.27	0.54	93.4
#4	PET fabric	12.67	12.23	52.74	22.36	1.15	127.5
#5		27.54	46.09	0.07	26.29	1.75	125.2
#6		3.00	27.61	44.86	24.53	1.09	130.3

Zhang et al. observed that the F/C ratio decreases with the increasing sputtering power, and increases with the increasing working pressure [20], the F/C ratio is affected by the sputtering power and working pressure significantly. the influence of substrate temperature on the properties of fluorocarbon films was mainly studied. Table 2 shows that the F/C ratio decreases with the increasing substrate temperature on PET plate, but an opposite result was obtained on the fabric substrate.

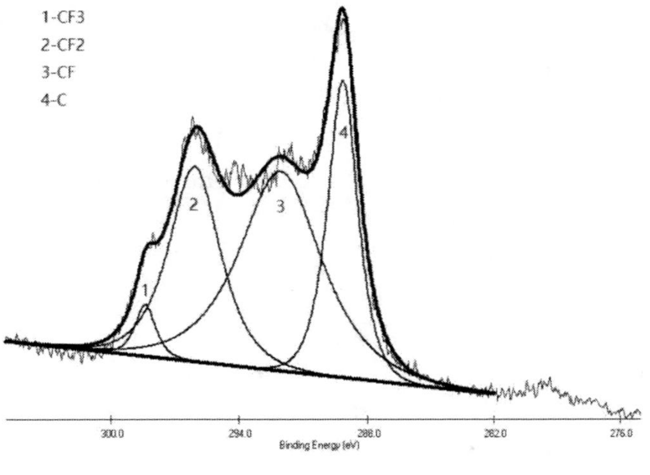

Fig. 2. A typical C1s spectrum of fluorocarbon film

In general, the smaller the F/C ratio, the weaker hydrophobicity of the film is [21]. Because fluorine atoms are hydrophobic groups, less fluorine atoms lead to a smaller F/C ratio, a smaller static contact angle and weaker hydrophobicity. This is true according to the results for Samples #1 - #3, in which the FC films are coated on PET plates. The solid surface is hydrophobic when the static contact angle is larger than 90 degrees, and in this case, the water drop is easy to be moved away the solid surface but not easy to wet the solid. It is slightly increased in the static contact angle of the coated PET plates because there was just particle clusters but no film formed on the surface shown in Fig.s 1 (a), (b) and (c). It seems that the substrate temperature has a slight influence on the F/C ratio and also the static contact angle.

On the contrary, the smaller the F/C ratio, the bigger the static contact angle of the coated fabric Sample #4 - #6 is. After sputtering for 120 min, the static contact angle turns to a value above 120 degrees. It means that the sample's surface changes by sputtering and the deposited films exhibit high hydrophobic properties. The F/C atoms were deposited on the surface and the surface's morphology and roughness change accordingly. At the same time, yarn weaving structure, holes, gaps and fiber microstructure all have impacts on the values of static contact angle.

4 Conclusion

The FC films were successfully coated on the PET substrate by RF magnetron sputtering. The composition of the films is made up of four components -CF3, -CF2-, CF- and -C-. The results indicate that the proportions of the four components and surface morphology of the films change with the different sputtering conditions, leading to changes in the static contact angle. The hydrophobicity of fabrics were improved after coating the FC films.

References

1. X. Yuan, Q. Wei, D. Chen, W. Xu, Text. Res. J. **86**(2016)
2. Y. Wu, L. Zhang, G. Min, H. Yu, B. Gao, H. Liu, S. Xing, T. Pang, Appl. Surf. Sci. **384**(2016)
3. S. Shahidi, M. Ghoranneviss, Clo. and Text. Res. J. **34**(2015)
4. Y. Chen, C. Hsu, J. He, Surf. and Coat. Tech. **232**(2013)
5. O. Baghriche, J. Kiwi, C. Pulgarin, R. Sanjinés, J. Photochem. Photobiol., A, **229**(2012)
6. P. Osorio-Vargas, R. Sanjines, C. Ruales, C. Castro, C. Pulgarin, A.J. Rengifo-Herrera, J.C. Lavanchy, J. Kiwi, J. Photochem. Photobiol., A, **220**(2011)
7. D. Depla, S. Segers, W. Leroy, T. Van Hove, M. Van Parys, Text. Res. J. **81**(2011)
8. J. Yip, S. Jiang, C. Wong, Surf. Coat. Technol., **204**(2009)
9. D. Hegemann, M.M. Hossain, D.J. Balazs, Prog. Org. Coat. **58**(2007)
10. S. Seino, S. Sasaki, T. Owashi, K. Oya, S. Iwamori, Vacuum, **111**(2015)
11. Y. Suzuki, H. Fu, Y. Abe, M. Kawamura, Vacuum, **87**(2013)
12. S. Iwamori, K. Noda, Mater. Lett. **66**(2012)
13. O. Kylián, M. Drábik, O. Polonskyi, J. Čechvala, A. Artemenko, I. Gordeev, A. Choukourov, I. Matolínová, D. Slavínská, H. Biederman, Thin Solid Films, **519**(2011)
14. S. Iwamori, N. Hasegawa, A. Uemura, T. Tanabe, I. Nishiyama, Vacuum, **84**(2009)
15. X. Sun, D. Li, B. Liu, Y. Zhang, Z. Ma, Surf. Coat. Technol., **201**(2007)
16. L.L.Meng, Q.F.Wei, X.M.Huang, F.F.Xu, Material Review, (2012)
17. F.B.Yu, X.H.Xia, Z.L.Yu, Q.J.Geng, Insulating Materials, (2008)
18. J.H.Hong, H.B.Wang, Q.F.Wei, Material Review, (2006)
19. H. Qi, Y. Zhang, J. Di, W. Du, Surface and Coatings Technology, **201**(2007)

20. Y. Zhang, Q. Ji, R. Liu, Z. Liu, H. Qi, Surf. Coat. Technol., **203**(2009)
21. J.S. Fang, C.S. Lin, Y.Y. Huang, T.S. Chin, J. Electron Mater, **44**(2015)

Preparation, and Luminescence Properties of SiO_2@Sm(MABA-Si)phen Core-Shell Structure Nanometer Composite

Li-Na Feng[1], *Wen-Xian* Li[1] * and *Jin-Rong* Bao[1]

[1] College of Chemistry and Chemical Engineering, Inner Mongolia University, Hohhot 010021, People's Republic of China

Abstract. A novel ternary samarium complex was prepared using $HOOCC_6H_4N(CONH(CH_2)_3Si\text{-}(OCH_2CH_3)_3)_2$ (MABA-Si) as first ligand, and phen as second ligand. The corresponding SiO2@Sm(MABA-Si)phen core-shell structure nanometer composite was synthesized as well, and the silica spheres was the core, and the ternary samarium complex was the shell layer. The ternary samarium complex has been characterized by element analysis, molar conductivity and IR spectra. The results show that the chemical formula of the complex is $Sm(MABA\text{-}Si)(phen)_2(ClO_4)_3\cdot 2H_2O$. The fluorescent spectra illustrat that the luminescence properties of the samarium complex are superior. The core-shell structure of SiO_2@Sm(MABA-Si)phen nanometer composite is characterized by SEM, TEM and IR spectra. The SiO_2@Sm(MABA-Si)phen core-shell structure composites exhibit stronger emission intensity than the ternary samarium complex. The fluorescence lifetime of the complex and core-shell structure composite is measured as well.

1. Introduction

The core-shell nanometer composite is a novel orderly assembled structure material. It can be formed by chemical bonds or physical adsorption. Therefore, it is a higher level of nanometer composite structure material. The core-shell nanometer composite differ from the unary substance because of its unique property, and it was intensely studied in the past decades [1-2]. By coating with a layer of other material, the surface of nanometer particle can be changed So the chemical and optical properties of nanometer particles can be greatly improved. Hence, the core-shell structure composite has the advantages that the unary material doesn't have.

Our work is to prepare the core-shell structure materials. SiO_2 Sphere is the core, and the rare earth organic complex is the shell, and they were connected together by silane coupling agent. By using silane coupling method, the core-shell materials were formed through chemical bond linkages which were formed between the SiO_2 core and outer material. In this way, the core-shell structure material was stable and the chemical bond was

* Corresponding author: nmglwx@163.com

not easy to break, so the morphology of core-shell material was easy to control. In this kind of core-shell structure materials, rare earth organic complex was used as cladding layer, and the thickness was magnitude nanometer. So the dosage of rare earth organic complexes can be reduced. As a result, the cost of materials can be greatly decreased. At the same time, compared with the hybrid materials [3,6], rare earth organic complex can be fixed by SiO_2 core easily, and the luminous efficiency of core-shell structure materials can be improved greatly.

In this paper, a novel ternary samarium complex $Sm(MABA\text{-}Si)phen_2(ClO_4)_3{\cdot}2H_2O$ has been synthesized for the first time. The SiO_2@Sm(MABA-Si)phen core–shell structure nanometer composite has been prepared as well. The samarum ions and SiO_2 core were connected together through MABA-Si silane coupling agent. The SiO_2@Sm(MABA-Si)phen core-shell structure composite exhibited stronger emission intensity than the ternary samarium complex.

2. The experiment

2.1 Materials

All chemicals were analytical grade and were used as received without further purification. Sm_2O_3 (purity > 99.99 %) were all supplied by Beijing Chemical Reagent Company. The Sm(III) perchlorate was prepared from dissolving in 2 $mol{\cdot}L^{-1}$ $HClO_4$, and then evaporated and dried in vacuum. phen, Tetraethoxysilane (TEOS, Aldrich), 3-(triethoxysilyl)-propyl isocyanate (96%, TEPIC, Aldrich).

2.2 Physical measurements

Elemental analysis was carried out on a HANAU analyzer. Conductivity measurement was made by using a 1×10^{-3}mol/L solution in dimethylformamide on a DDS-11D conductivity meter. The dysprosium content of the complex was determined by EDTA titration using xylenol-orange as indicator. The infrared spectra was determined by the KBr pressed disc method on a Nicolet NEXUS-670 FT-IR spectrophotometer. The 1HNMR spectra was determined in the DMSO-d_6 by Bruker AC-500 nuclear magnetic resonance spectrophotometer. Luminescence properties were determined on a FLS980 fluorescence photometer with the slit width of 1nm. Fluorescent decay curves were recorded by FLS980 Combined Steady State and Lifetime Spectrometer. The microstructure were recorded by scanning the electronic microscope (Hitachi S-4800), transmission electron microscope （FEI Tecnai F20)

2.3 Synthesis of the ternary samarium complex

The ligand MABA-Si was synthesized according to the method of the literature [7,8]. The product has been characterized by 1HNMR spectra. 1HNMR data: δ0.56ppm(4H), δ1.04-1.51ppm(18H), δ2.93ppm(4H), δ3.33-3.47ppm(12H), δ3.73ppm(4H), δ7.43-7.57ppm(2H) and δ12.4 ppm(1H). Yield: 40%. mp: 46~48°C.

1 mmol MABA-Si and 2 mmol phen anhydrous ethanol was added into 1 mmol $Sm(ClO_4)_3{\cdot}nH_2O$ anhydrous ethanol solution drop by drop under the magnetic stirring for 2 h. A white precipitate was formed. The resulting white solid precipitates were filtered, washed several times with anhydrous alcohol. (yield>80%)

2.4 Synthesis of the core-shell structure nanometer composite SiO_2@Sm(MABA-Si)phen

2.4.1 Synthesis of SiO_2@MABA-Si

Firstly, mono-disperse SiO_2 were synthesized by the well-known Stöber[9] method. Then 0.1 g SiO_2 was dissolved in 20 mL mixture solution of anhydrous alcohol and distilled water under stirring at 12 hours. 0.2 g MABA-Si was dissolved in 15 ml anhydrous alcohol, adding drop-wise to the activated SiO_2 microspheres solution that mentioned above under stirring for 10 h. A white precipitate was centrifuged separation, and washed by distilled water and anhydrous ethanol several times, then dried at 40 °C.

2.4.2 Synthesis of core-shell structure nanometer SiO_2@Sm(MABA-Si)phen

SiO_2@MABA-Si was dispersed in 10 mL anhydrous ethanol. The molar ratio of Sm^{3+}: MABA-Si: phen equals 1:1:2, the phen was dissolved in 5 ml anhydrous alcohol and was added into the SiO_2 @MABA-Si anhydrous ethanol drop by drop, The appropriate amount samaium perchlorate was dissolved in 5 ml anhydrous ethanol and then added into the anhydrous ethanol solution of phen and SiO_2 @MABA-Si drop by drop. The mixture was stirred 5 h, centrifuged separation. The product was washed by anhydrous ethanol several times and then dried. A white precipitate was obtained.

3 Results and discussions

3.1 The composition and Properties of the complex

The elemental analysis data of ternary samarium complex was shown as follows: C, 39.54; N, 6.33; H, 4.45 (Anal. Calcd: C, 40.33; N, 6.47; H, 4.63), and it was conformed to $Sm(MABA\text{-}Si)(phen)_2(ClO_4)_3 \cdot 2H_2O$. The complex was white powder, and it was stable in atmospheric condition and soluble in DMF and DMSO. The result of molar conductivity value (149 $S \cdot cm^2 \cdot mol^{-1}$) of the complexes in DMF was formulated 1:2 electrolytes [10].

3.2 SEM and TEM

In our work， the core-shell structure composite was formed by hydrolysis between the ethoxy of MABA-Si and hydroxyl groups of the SiO_2 surface. Then the carboxyl oxygen atoms of MABA-Si groups and the double nitrogen of phen can be coordinated with Sm^{3+} ions. As a result, the SiO_2 cores and Sm ions were connected together through the "molecular bridge", the MABA-Si. The peripheral of SiO_2 cores was coated by Sm(MABA-Si)L complex successfully. The core-shell structure was observed by SEM, TEM and Infrared spectra.

SEM and TEM images of SiO_2 microspheres and core-shell structure composites were shown in the Fig 1,2. Fig.1. could be seen that the SiO_2 had the regular micro-structure, smooth surface and good dispersity. The average particle size of SiO_2 spherical was about 400nm. Core-shell structure nanometer SiO_2@Sm(MABA-Si)phen can be seen in Fig.2. Resulted from TEM photographs, the interface of the core-shell structure micro-spheres show that the surface is much rougher than the pure SiO_2 micro-spheres. The thickness of the gray shell layer of the composites was about 20 nm.

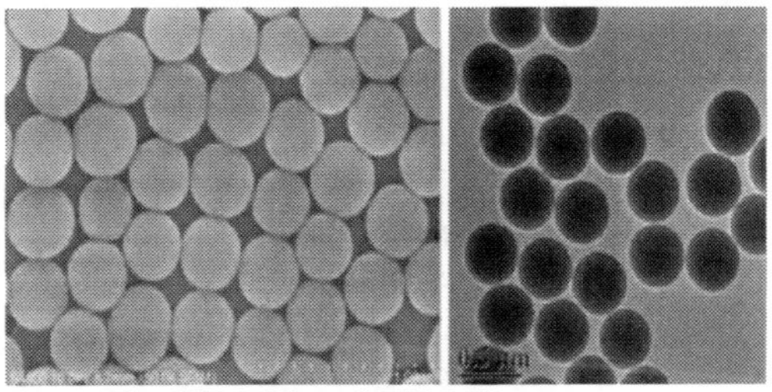

Fig .1. The SEM (L) and TEM(R) photographs of 400nm SiO_2

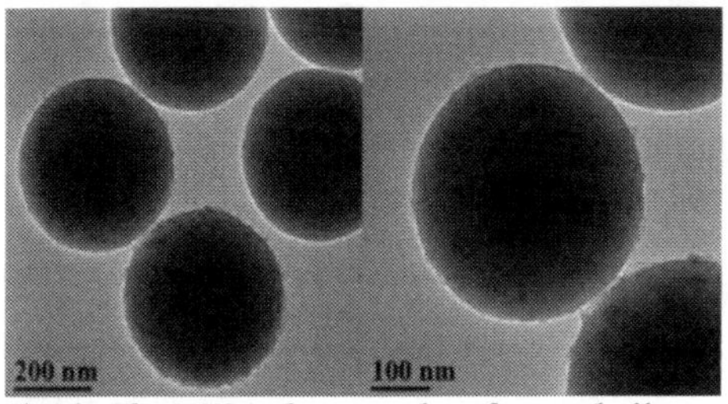

Fig .2. The TEM photographs of core-shell structures SiO_2@ Sm(MABA-Si)phen

3.3 Infrared spectra

Based on Fig.3a-c,For MABA-Si, the $\nu_{C=O(COOH)}$, $\nu_{C-OH(COOH)}$ and $\delta_{O-H(COOH)}$ appeared at 1700 cm^{-1}, 1447 cm^{-1}, and 1316 cm^{-1}. After coordinating with Sm^{3+}, they shifted to lower wave number which appeared at 1682 cm^{-1}, 1398 cm^{-1}, 1242 cm^{-1}. The results showed that rare earth samarium ion bonded with MABA-Si through the COOH group [11,13]. For phen, $\nu_{C=N}$ shifted to 1521 cm^{-1}, and the δ_{C-H} shifted to 716 cm^{-1} and 848 cm^{-1} respectively, comparing with pure phen (the $\nu_{C=N}$ appeared at 1587 cm^{-1}, and δ_{C-H} appeared at 735 cm^{-1} and 851 cm^{-1} respectively). The results showed that rare earth samarium ion bonded with phen through the double nitrogen atoms. In the samarium complex, there were four perchlorate group absorptions at 1142 cm^{-1}, 1115 cm^{-1}, 1079 cm^{-1}, and 625 cm^{-1}. Combined the molar conductivity, one perchlorate group bonded with the Sm(III) ion by the oxygen atoms, and the coordination was bidentate. The IR spectrum of SiO_2 cores, MABA-Si, phen, SiO_2@Sm(MABA-Si)phen can be seen in Fig.4a-d. In the IR spectrum of MABA-Si and the core-shell structure nanometer composite, the $\nu_{C=O(COOH)}$ appeared at 1700 cm^{-1}, 1656 cm^{-1} , respectively. The ν_{-CONH-} of MABA-Si appeared at 1639cm^{-1} and 1558 cm-1 respectively. However in the core-shell composite, the ν_{-CONH-} group shifted to lower wave number which appeared at 1619 cm^{-1} and 1556 cm^{-1} respectively. Similarly, the $\nu_{C=N}$ of phen group and the core–shell composite appeared at 1587 cm^{-1}, 1522 cm^{-1}. These results suggested that the Sm-MABASi-phen complexes were formed in the shell layer of core–shell composites.

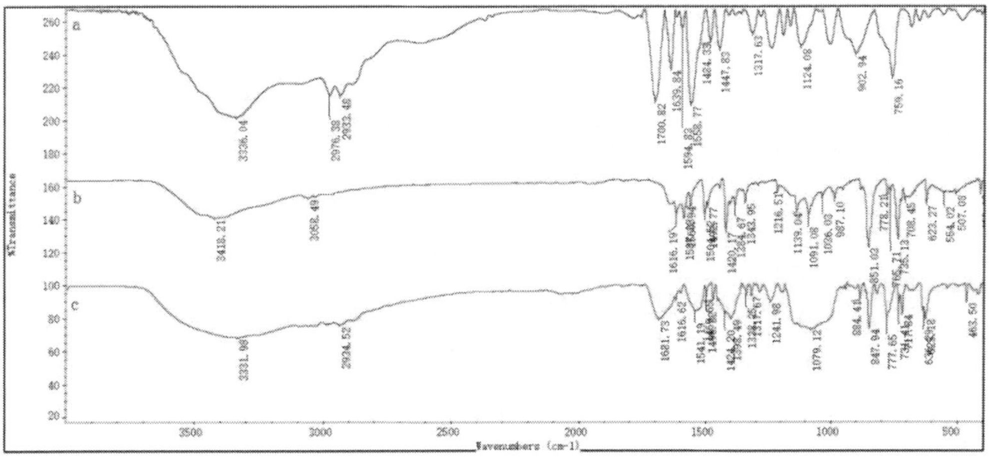

Fig. 3. The IR Spectra of (a). MABA-Si ; (b). phen; (c). $Sm(MABA-Si)(phen)_2(ClO_4)_3 \cdot 2H_2O$

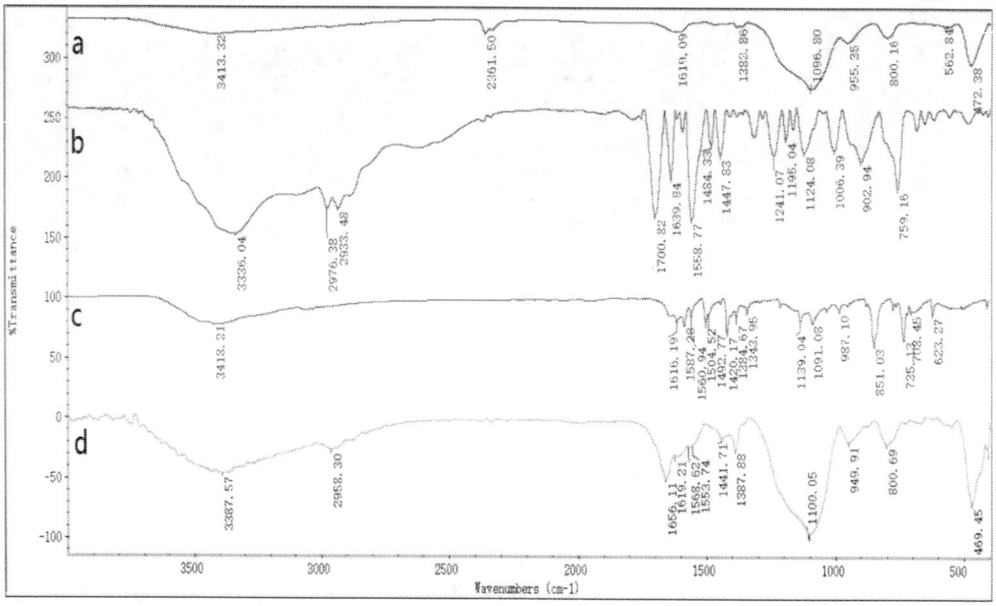

Fig. 4. The IR Spectra of (a). SiO_2 ; (b). MABA-Si; (c).phen; (d).SiO_2@Sm(MABA-Si)phen

3.4 Fluorescence spectra

The fluorescence excitation and emission spectrum of core shell structure composites and ternary samarium complex were measured in solid state at room temperature. The results were shown in Fig 5a,b There were three emission peaks, which were attributed to the characteristic transitions of Sm(III). The three peaks correspond to $^4G_{5/2}$-$^6H_{5/2}$, $^6H_{7/2}$, $^6H_{9/2}$ and the peaks were appeared at 561 nm, 596 nm, and 643 nm respectively. The strongest fluorescence emission intensity of core–shell structure composites and ternary samarium complex were 860847a.u. and 295259a.u. The fluorescence emission intensity of core-shell structure composites and ternary samarium complex was 2.92 times as great as the ternary complex system.

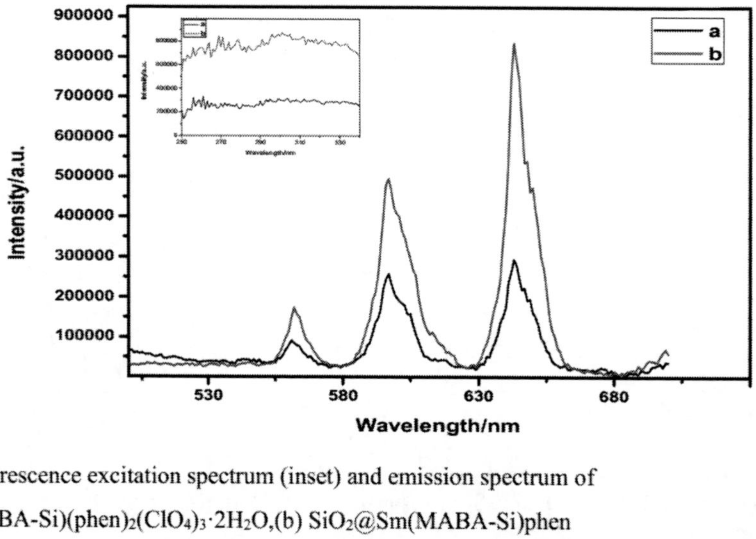

Fig. 5. Fluorescence excitation spectrum (inset) and emission spectrum of

(a) Sm(MABA-Si)$(phen)_2(ClO_4)_3 \cdot 2H_2O$,(b) SiO_2@Sm(MABA-Si)phen

3.5 The fluorescence decay curve

The fluorescence decay curve of the ternary samarium complex and the core-shell structure composites was also measured. Fig.6-7 showed the fluorescence decay and double exponential fit curve of the $Sm(MABA\text{-}Si)(phen)_2(ClO_4)_3{\cdot}2H_2O$ and SiO_2@Sm(MABA-Si)phen respectively. The lifetime of the Sm(III) ion could be obtained from the Eq.

$$I_{(t)} = I_0 + A_1 \exp(-t_1/\tau_1) + A_2 \exp(-t_2/\tau_2) \tag{1}$$

$$<\tau> = (A_1\tau_1^2 + A_2\tau_2^2)/(A_1\tau_1 + A_2\tau_2) \tag{2}$$

Where $I_{(t)}$ was the fluorescence intensity varying with time t, and τ_1 and τ_2 were lifetime. The corresponding data were calculated. As the result, the lifetime of the ternary samarium complex was 31.27μs. Moreover, the lifetime of the core-shell composites was 39.58μs.

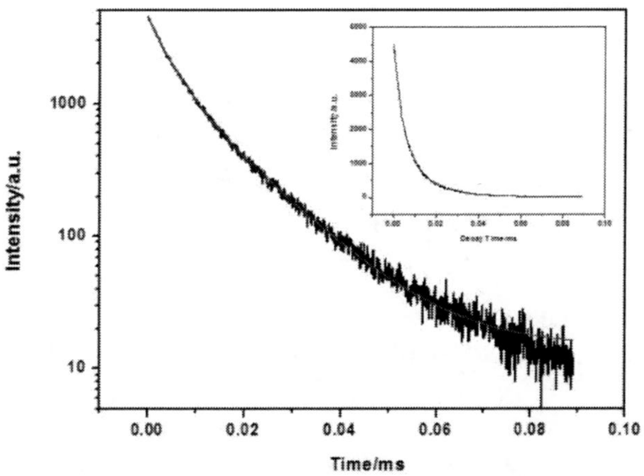

Fig. 6. Decay Curve (R)and Fit Curve (L) of $Sm(MABA\text{-}Si)(phen)_2(ClO_4)_3{\cdot}2H_2O$

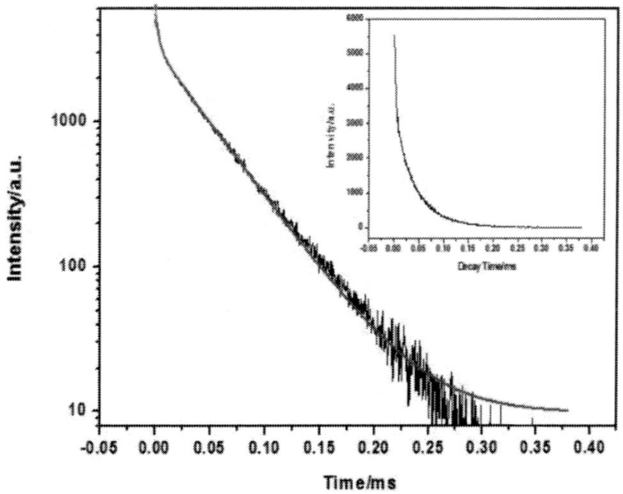

Fig. 7. Decay Curve (R) and Fit Curve (L) of SiO_2@Sm(MABA-Si)phen

4 Conclusions

A novel ternary samarium complex Sm(MABA-Si)(phen)$_2$(ClO$_4$)$_3$·2H$_2$O and the corresponding core-shell structure composites SiO$_2$@Sm(MABA-Si)phen were prepared and characterized. Luminescence property was studied by the fluorescence spectrum. The results showed that they had excellent luminescent property. The fluorescence emission intensity of the core-shell structure composites system was 2.92 times as great as the corresponding ternary complex system. The result above showed that SiO$_2$ core and ligands played a mutually synergistic effect in the energy transfer process from the ligands to Sm(III) ion. The evidence illustrated that the fluorescence emission intensity of material can be greatly enhanced by the formation of core-shell structure. The SiO$_2$@Sm(MABA-Si)phen core–shell structure nanometer composites can be applied in luminescent materials. So this approach gave an applicable new choice for future study of novel luminescent materials. These studies can help us find more about samarium luminescent materials, as well as develop the application of rare earth elements samarium.

This work is supported by the financial supports from the Major projects of Natural Science Foundations of Inner Mongolia Science Foundation (2015ZD01) and supported by the Natural Science Foundations of Inner Mongolia Science Foundation (2015MS0502)

References

1. Y L Liu, G Cheng, Z G W, etc. Mater Lett.[J], **97**: 187-190 (2013)
2. X Y Huang. J Alloy Compd [J], **628**:240-244 (2015)
3. Y Li , B Yan. Journal of Solid State Chemistry, **181**:1032-1039 (2008)
4. Z Y Yan, B Yan. Inorganic Chemistry Communications, **47**:96-98 (2014)
5. Y Ban, C Wang. Inorganic Chemistry Communications, **14**:1494-1497 (2011)
6. X M Guo, H D Guo, L S Fu, Zhang H J * etal. Microporous and Mesoporous Materials, **119**:252–258 (2009)
7. Q M Wang, B Yan. J Mater Chem.[J], **14**:2450-2454 (2004)
8. H R Li, J Lin, H J Zhang, etc. Chem. Mater[J], **14**:3651-3655 (2002)
9. G L Davies, A Barry, K Yurii. Gunko. Chem. Phys. Lett. [J], **468**:239-244 (2009)
10. W Geary. J Coord Chem. Rev.[J], **7**:81-121 (1971)
11. L J Yuan, M C Yin, E T Yuan, etc. Inorg Chem. Acta, **375**: 89-94 (2004)
12. C Qin, X L Wang , E B Wang, etc. Inorg Chem.[J], **44**: 7122-7129 (2005)
13. V A Petrov, W J Marshall, V V Grushin. Chem. Comm. [J], **5**: 520-521 (2002)

Synthesis, and Characteristic Luminescence of Core-Shell SiO_2@Eu(MABA-Si)·(phen) Microspheres

Yang-Yang Ma[1], *Wen-Xian* Li[1*], and *Yu-Shan* Zheng[2]

[1]College of Chemistry and Chemical Engineering, Inner Mongolia University, Hohhot 010021, People's Republic of China
[2]Inner Mongolia Autonomous Region food inspection test center, Hohhot, 010021,People's Republic of China

Abstract. Monodispersed SiO_2@Eu(MABA-Si)·(phen) core-shell microspheres were synthesized by silane coupling agent method. The core-shell microspheres was used the europium complex Eu(MABA-Si)·$(phen)_2$·$(ClO_4)_3$·$2H_2O$ as shell and SiO_2 as the inorganic core. The europium complex shell has grafted to silica surface through forming a Si-O-Si bond. The europium complex was synthesized by $HOOCC_6H_4N$- $(CONH(CH_2)_3Si(OCH_2CH_3)_3)_2$ (denoted as MABA-Si) and phen coordinated europium perchlorate. The europium complex has been characterized by element analysis, molar conductivity, ^{1}HNMR and IR. The TEM and SEM showed that the diameter of SiO_2 core was about 400 nm and the thickness of the europium complex shell was about 20 nm. Fluorescence spectra illustrated that the core-shell microspheres have stronger fluorescence intensity than the europium complex which was 1.87 times.

1 Introduction

Core-shell nanometer microspheres are a kind of nanomaterial with a core-shell structure that can be widely applied in making functional fluorescent materials, magnetic materials, electronics, medicine, and so on. By coating a layer of organic alkoxysilane complexes, the luminescence properties of core-shell materials can be changed *including* luminescence intensity, fluorescence quantum efficiency, stability and fluorescence lifetime [1-3].

Silica probably is the most popular inorganic material because of its good mono-disperse, flexibility of controlling the particle size, low cost, low toxicity, and more mechanical stability. Owing to the external hydrogen bonds of its surface, SiO_2 is easy to bond with different materials through chemical bonds or electrostatic interactions [4-6]. At present, silane coupling agent is used to prepare the coating structure nanomaterials that was generally described as $Y(CH_2)nSiX_3$. Y group is the organic functional group that can combine with the rare earth ions. X represents the hydroxyl groups which combined with

*Corresponding author: nmglwx@163.com

Si-OH of the surface of SiO_2. As a result, RE^{3+} and SiO_2 can be incorporated through silane coupling agent which plays a role of the "molecular bridge".

In this article, we have controlled prepared ternary europium complex Eu(MABA-Si)·(phen)$_2$·$(ClO_4)_3$·$2H_2O$. The MABA-Si is a kind of silane coupling agent, which can be easily coordinated with Eu^{3+} ions through the carboxyl oxygen atoms of MABA-Si groups, and the second ligand phen also can be coordinated with Eu^{3+}. Furthermore, the monodispersed SiO_2@Eu(MABA-Si)·(phen) core-shell microspheres synthesized by the europium complex grafting to silica surface through forming a Si-O-Si bond. The luminescent properties of europium complex and core-shell microspheres SiO_2@Eu(MABA-Si)·(phen) was discussed in detail.

2 The experiment

2.1 Materials.

Eu_2O_3(99.99%), $HClO_4$(1 mol·L^{-1}), phen, m-aminobenzoic acid, TEOS, 3-(Triethoxysilyl)-propyl isocyanate (96%, TEPIC, Aldrich), All other reagents were analytically grade.

2.2 Physical measurements.

Elemental analysis was carried on PF-2400 Elemental Analyzer. Conductivity was measured by DDS-11D Conductivity Meter. The europium ion content was measured by EDTA titration, Xylenol-Orange as indicator. The IR was determined by Nicolet NEXUS-670 FT-IR spectrophotometer. The ^{1}HNMR spectra was determined in the DMSO-d_6 by Bruker AC-500 nuclear magnetic resonance spectrophotometer.The morphology of the products was observed by Scanning Electronic Microscope (Hitachi S-4800) and Transmission Electronic Microscope (FEI Tecnai F20). Fluorescence properties were determined via FLS980 (slit width: 1 nm). Fluorescent decay curves were recorded by FLS980 Combined Steady State and Lifetime Spectrometer.

2.3 Preparation of the europium ternary complex.

The ligand MABA-Si was synthesized according to the method of the literature [7]. 1 mmol MABA-Si and phen were added into 1 mmol $Eu(ClO_4)_3 \cdot nH_2O$, anhydrous ethanol solution drop by drop under the magnetic stirring. White precipitate was obtained by washing for several times with anhydrous ethanol and distilled water.

2.4 The synthesis of core-shell SiO_2@Eu(MABA-Si)·(phen) microspheres.

SiO_2 was synthesized through the so-called Stöber [8] method. 0.2 g MABA-Si was added into 15 ml SiO_2 anhydrous ethanol solution under magnetic stirring for 12h. Then appropriate concentration of europium perchlorate and phen were added into above solution. The white precipitate core-shell microspheres SiO_2@Eu(MABA-Si)·(phen) were formed by washing for several times with anhydrous ethanol and distilled water.

3 Results and discussions

3.1 The composition of the europium complex.

The composition and structure of the europium complex have been identified by element analysis, molar conductivity, ^{1}HNMR and IR. The element analysis showed that Anal. Calcd for $C_{28}H_{49}N_3O_{10}Si_2$ (MABA-Si): C, 51.35%; H, 7.76%; N, 6.66%, found: C, 50.80 %; H, 7.82%; N, 6.55%. The results demonstrated that the experimental value was close to the theoretical value. The molar conductivity value was 167 $S \cdot cm^2 \cdot mol^{-1}$, (DMF as solvent). The ^{1}HNMR (DMSO) spectra of europium complex's chemical shift appeared at δ0.56 ppm, δ1.04-1.51 ppm, δ2.93 ppm, δ3.33-3.47 ppm, δ3.73 ppm, δ7.43-7.57 ppm and δ12.4 ppm. It showed that MABA-Si synthesized successfully. Fig. 1 displayed the IR spectra of MABA-Si, phen and Eu(MABA-Si)·(phen)·$(ClO_4)_3 \cdot 2H_2O$. The peak located at 1700, 1417, and 1314 cm^{-1} were belonged to the $\nu_{C=O}$, ν_{C-OH} and δ_{O-H} of carbonyl, which prove the presence of MABA-Si. Compared with MABA-Si, the europium complex not only exhibit similar infrared absorption bands as MABA-Si but also the peak at 1527, 716 and 848 cm^{-1} range in phen with red shift, indicating that europium ion was bonded with MABA-Si and phen through the carbonyl group and double nitrogen atoms [9]. It was found that ClO_4^- absorption band around 1145, 1115, 1083, and 625 cm^{-1}, which were assigned C_{2v} symmetry of the perchlorate group [10]. The above analysis shows that the composition of the complex was Eu(MABA-Si)·$(phen)_2 \cdot (ClO_4)_3 \cdot 2H_2O$.

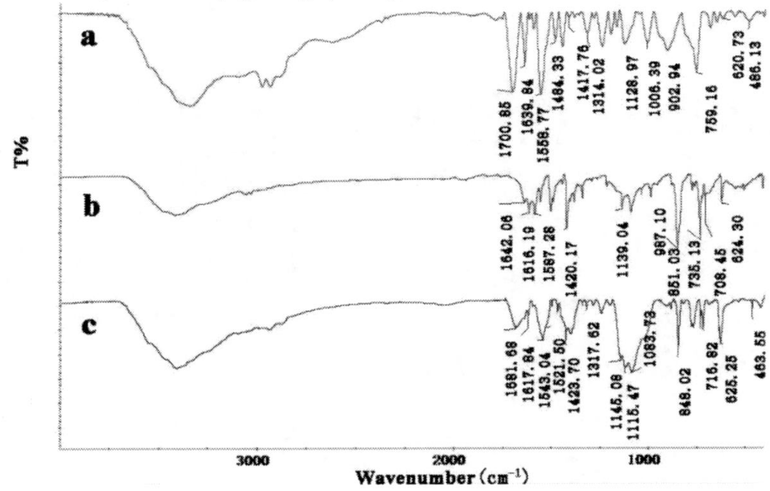

Fig. 1. IR Spectra of a. MABA-Si; b. phen; c. Eu(MABA-Si)·$(phen)_2 \cdot (ClO_4)_3 \cdot 2H_2O$

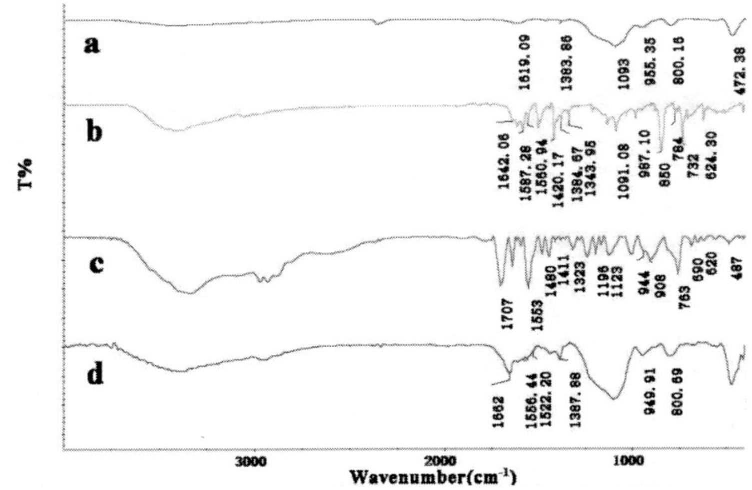

Fig. 2. IR Spectra of a.SiO_2; b. MABA-Si; c. phen; d. SiO_2@Eu(MABA-Si)·(phen)

3.2 IR spectra analysis of core-shell microspheres

Fig. 2 displays the IR spectra of SiO_2, MABA-Si, phen, and SiO_2@Eu(MABA-Si)·(phen). The absorption band Si-O-Si was appeared at 1096 and 800 cm^{-1}(Fig. 2a and 2d), which confirmed that the europium complex shell has grafted to silica surface through forming a Si-O-Si bond. In Fig. 2b, 2c and 2d, it was similar to above complex that the stretching frequency red shifted to 1656, 1619, 1553 cm^{-1} (MABA-Si) and 1568 cm^{-1} (phen). Compared with 1700, 1639 and 1558 cm^{-1} (MABA-Si) and 1587 cm^{-1} (phen). The above information indicated that europium ion coordinated with through C=O oxygen atom of MABA-Si and double nitrogen of phen.

3.3 The SEM and TEM analysis of core-shell microspheres.

The morphology was revealed by SEM and TEM images (Fig. 3a and 3b). The d3iameter of SiO_2 core was about 400 nm. The TEM images of the core-shell SiO_2@Eu(MABA-Si)·(phen) microspheres were presented in Fig. 3c and 3d, the thickness of coating layer was about 20 nm. The core-shell microspheres showed rougher surface than pure SiO_2 particles. In addition, characteristic peaks of europium was found in the EDX spectrum of core-shell microspheres, which confirmed that Eu(MABA-Si)·(phen) exist in the cladding layer of core-shell microspheres (Fig. 4).

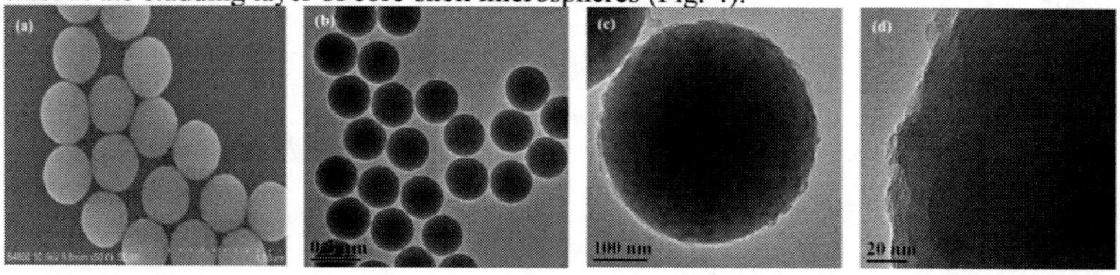

Fig. 3. a, the SEM of SiO_2; b, TEM of SiO_2; c and d, SiO_2@Eu(MABA-Si)·(phen)

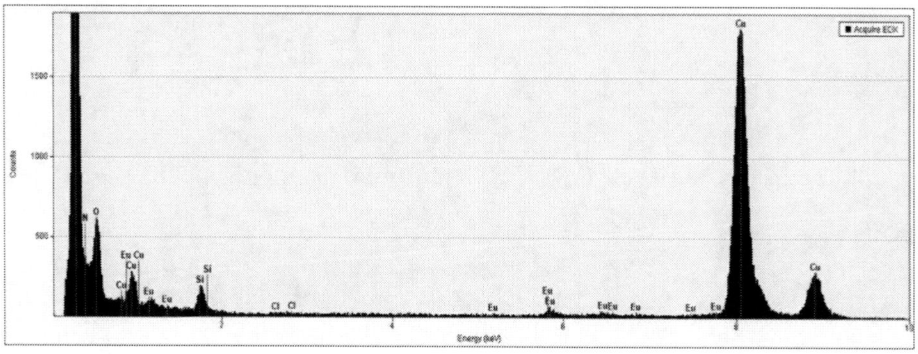

Fig. 4. EDX spectrum of core-shell microsphere

3.4 Luminescent properties.

The emission spectra of the europium complex (Fig. 6b) and core-shell microsopheres SiO_2@Eu(MABA-Si)·(phen) (Fig. 6a) were recorded under excitation wavelength at λex=272 nm that exhibited the characteristic transitions $^5D_0 \rightarrow {}^7F_J$ (J=0-4) of Eu^{3+}, and the peaks are appeared at 579, 593, 612, 652 and 699 nm respectively. The strongest emission peak appears at 612 nm, which was assigned to $^5D_0 \rightarrow {}^7F_2$ transition of Eu^{3+}. The emission intensity of the core-shell microspheres (36282456 a.u.) stronger than the europium complex (19425658 a.u). The fluorescent lifetime value of the europium complex and core-shell microspheres (Fig. 5a) were calculated by the double exponential mode. The core-shell microspheres lifetime was 1009.67 μs which langer than 389.55 μs of the europium complex (Fig. 5b).

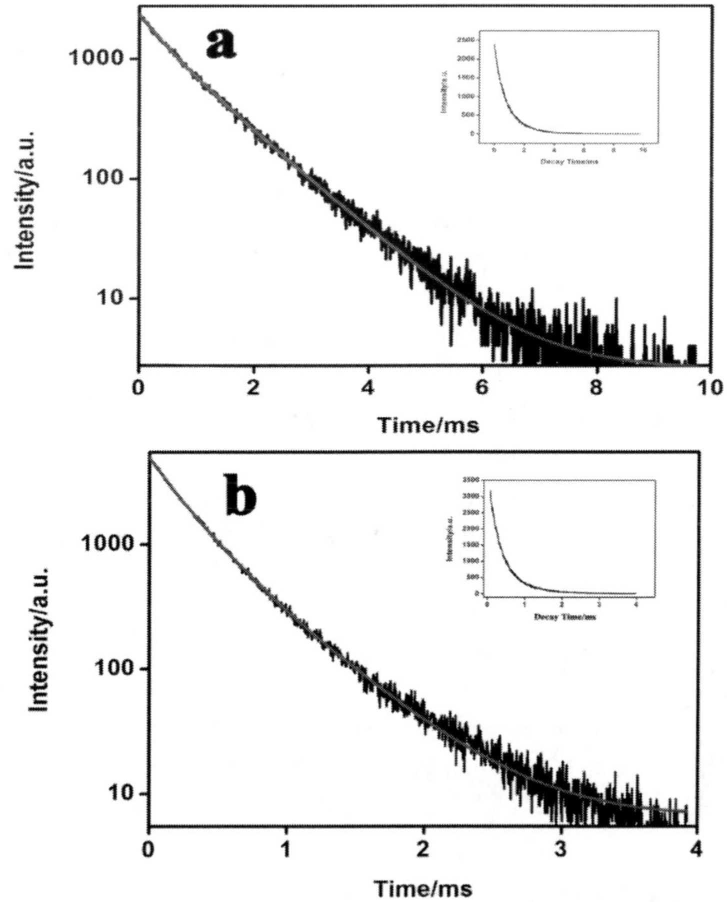

Fig. 5. Decay Curve and Fit Curve of core-shell microsphere (a) and europium complex (b)

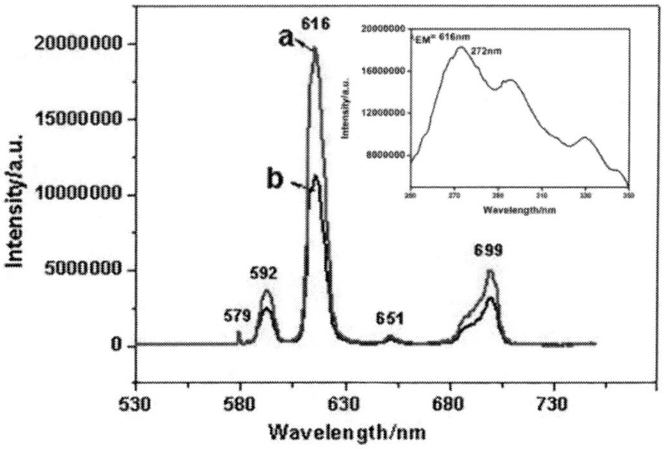

Fig. 6. excitation, emission spectrum of core-shell microspheres (a) and europium complex (b)

4 Conclusions

The novel europium complex of Eu(MABA-Si)(phen)$_2$·(ClO$_4$)$_3$·2H$_2$O and core-shell SiO$_2$@Eu(MABA-Si)(phen) microspheres were prepared. The core-shell microspheres was used the europium complex as shell and SiO$_2$ as the inorganic core. The europium complex shell has grafted to silica surface through forming a Si-O-Si bond. The emission intensity of core-shell microspheres was enhanced 1.87 times as great as the europium complex. And the core-shell microspheres have longer lifetime. The core-shell microspheres have a potential application as red luminescence nanomaterials, which will expand their applications in optical or electronic field.

This work is supported by the Major projects of Natural Science Foundations of Inner Mongolia Science Foundation (2015ZD01) and the Natural Science Foundations of Inner Mongolia Science Foundation (2015MS0502).

References

1. X. Li, O. Niitsoo, and A.Couzis, J Colloid Interface Sci **465** 333-341 (2016)
2. J. Zeng, et al. Anal Chim Acta **891** 269-276. (2015)
3. H.Zhang, D. Lin., G. Xu, J.Zheng, .Int. J. Hydrogen Energy **40** 1742-1751 (2015)
4. H. Chen, Y.B. Lang, Y.L. Zhang, et.al. J. Mater. Chem. C **3** 6314-6321 (2015)
5. T. Kurabayashi, T. Yamaki, T. Fukuda, et. al. Mol. Cryst. Liq. Cryst. **621** 136-142 (2015)
6. T. Fukuda, T. Kurabayashi, T. Yamaki, et. al. Nanotechnol. **16** 3235-3240. (2016)
7. Davies, G. L., Barry, A., Gun'Ko, Y. K, Chem. Phys. Lett. **468** 239-244 (2009)
8. Geary, W. J, Coord. Chem. Rev. **7** 81-122 (1971)
9. Liangjie,Yuan, et al, Inorg. Chim. Acta. **357** 89-94 (2004)
10. A. Balamurugan, M.L. Reddy, M. Jayakannan, J. Phys. Chem. B. **113** 14128-14138 (2009)

Luminescence Properties of SiO_2@Eu(phen-Si) (sulphoxide) Core-Shell Nanometer Composite

Yi-Lian Li[1], *Wen-Xian* Li[1*], *Yu-Shan* Zheng[1]

[1]College of Chemistry and Chemical Engineering, Inner Mongolia University Hohhot 010021, China
[2]Inner Mongolia Autonomous Region food inspection test center, Hohhot 010021, China

Abstract. A novel ternary europium complex Eu(Phen-Si)·$(H_2L)_2$·$(ClO_4)_3$·$6H_2O$ was prepared using (phen)-N-$(CONH(CH_2)_3Si(OCH_2CH_3)_3)_2$ (Phen-Si) as the first ligand and 2-carboxyphenyl carboxymethyl sulphoxide (H_2L) as the second ligand. The corresponding different thickness core-shell structure SiO_2@Eu(phen-Si)·L nano composites were synthesized, with silica spheres as core, ternary europium complex (Eu(Phen-Si)·$(H_2L)_2$·$(ClO_4)_3$) as shell. The ternary europium complex has been characterized by element analysis, molar conductivity, and IR spectra. The SiO_2@Eu(phen-Si)·L core-shell structure composite was characterized by TEM and IR spectra. Core-shell structure composites exhibited stronger fluorescence properties than that of the ternary europium complex. The emission intensity of the two core-shell composites was 1.79, 3.99 times stronger than the ternary europium complex, respectively. The fluorescence lifetime of the core-shell structure composites were longer than the ternary europium complex.

1 Introduction

The core-shell nano composite is a nanoparticle of novel structure. It is an orderly assembly structure, which can be constituted of a nanomaterial coated with another nanomaterial by chemical bonds or physical adsorption. Therefore, it is a higher level nano composite structure. Because of its unique property, differing from any of its unary substance, many core-shell nano composites are intensely studied in the past decades [1-3].

Coating with other materials, the surface of nanoparticle changes the structure of the layer, size, or composition. It has the advantages that unary material does not have and open up a new approach for the new material design. Because of its easily controlling particle size, low cost, security, and low toxicity, SiO_2 is the most widely used core-shell structure nanoparticles. As a hotspot in the field of nanometer materials, the study of SiO_2 as core materials was reported more and more, with the rare earth inorganic materials used most as the coating layer [4-7]. Instead, research of rare earth organic complexes as cladding layer was reported few.

Silane coupling agent is a kind of bi-functional organic matter, generally formulated as $Y(CH_2)nSiX_3$. Y represents the organic functional groups, such as amino, carboxyl and double nitrogen, etc. X represents the alkoxy. Y group can combine with rare earth ions. X

* Corresponding author: nmglwx@163.com

group can form Si-O-Si chemical bonds with hydroxyl groups of the SiO_2 through hydrolysis. As a result, rare earth ions and SiO_2 are connected together by silane coupling agent, as a "molecular bridge". Yan Bing and Zhang HongJie teams did a lot work in this area [8-9].

In this article, the phenanthroline of amino silane functionalization phen-Si, was used as a silane coupling agent. A novel ternary europium organic complex $Eu(Phen\text{-}Si)\cdot(H_2L)_2\cdot(ClO_4)_3\cdot 6H_2O$ has been synthesized for the first time, using Phen-Si as the first ligand, 2-carboxyphenyl carboxymethy sulphoxide(H_2L) as the second ligand. The SiO_2@Eu(phen-Si)·L core-shell structure nano composites have been prepared, using silica spheres as the core, the ternary europium organic complex $Eu(Phen\text{-}Si)\cdot(H_2L)_2\cdot(ClO_4)_3\cdot 6H_2O$ as the shell, denoted as SiO_2@Eu(phen-Si)·L.·Analytical and characterization tests were applied to the prepared complex and nano-particle composite.

2 The experiment

2.1 Materials

All chemicals were analytical grade and were used as received without further purification. Eu_2O_3 (purity > 99.99%) were all supplied by Beijing Chemical Reagent Company. The Eu(III) perchlorate was prepared from dissolving in 2 $mol\cdot L^{-1}$ $HClO_4$, and then evaporated and dried in vacuum. 2-carboxyphenyl carboxymethyl sulphoxide (H_2L), phen, Tetraethoxysilane (TEOS, Aldrich), 3-(triethoxysilyl)-propyl isocyanate (96%, TEPIC, Aldrich).

2.2 Physical measurements

Elemental analysis was carried out on a HANAU analyzer. Conductivity measurement was carried out on a DDS-11D conductivity meter. The europium content of the complex was determined by EDTA titration. The infrared spectra was determined on a Nicolet NEXUS-670 FT-IR spectrophotometer. Fluorescence properties were determined on a FLS920 fluorescence spectrometer. Microstructures were recorded by Scanning electronic microscope (Hitachi S-4800), and transmission electron microscope (FEI Tecnai F20).

2.3 Synthesis of the ligand phen-Si and the ternary rare earth complex

Phen-Si was synthesized according to the method of the literature [10-12]. Yield: 40%.Mp: 46~48°C.Anal.calcd. For $C_{32}H_{51}N_5O_8Si_2$(M=689g/mol): C, 56.23%; H, 7.31%; N, 10.21%, found: C, 55.73%; H, 7.43%; N, 10.16%. Then 1 mmol Phen-Si and 2 mmol H_2L were dissolved in ethanol. The solution of 1 mmol $Eu(ClO_4)_3\cdot nH_2O$ ethanol was added into the above mixed solution under stirring at 80℃ for 0.5 h. Finally, white precipitate was obtained by washing for several times with ethanol and water. (yield>90%).

2.4 Synthesis of the SiO_2@Eu(phen-Si)·L core-shell structure nano composites

Mono-disperse spherical SiO_2 was synthesized by Stöber method. The average diameter of SiO_2 was 250nm. Phen-Si was dissolved in 5 ml ethanol, adding to the solution which was a mixture contained SiO_2, 5 ml alcohol and 5 ml water, and the pH was set to 9 by an appropriate amount of ammonia. Then the mixture was stirred 2h at room temperature.

Finally, the core-shell structure nano composites were successfully synthesized. 1 ml SiO_2@(phen-Si) was dispersed in 20 ml ethanol. The molar ratio of SiO_2@(phen-Si): sulfoxide: Eu^{3+} was 1:2:1. The H_2L and Eu(III) perchlorate were added into the silica solution sequentially. The mixture was stirred 0.5 h at 80℃. Then a white precipitate was obtained.

3 Results and discussions

3.1 The composition the complex

Analytical data for the complex, presented in Table 1. The result of molar conductivity value of the complex in DMF was accordingly formulated as 1:2 electrolytes [13]. EDTA titration result showed that the content of the europium ion was 9.22%. According to the elemental analysis and EDTA titration data, the composition of the ternary complex was suggested as Eu(Phen-Si)·$(H_2L)_2$·$(ClO_4)_3$·$6H_2O$.

Table 1. Composition (%) and Molar Conductivities ($S\cdot cm^2\cdot mol^{-1}$) of Rare earth Complex (25°C)

Complex	C	N	H	RE	λ_m
Eu (Phen-Si)·$(H_2L)_2$·$(ClO_4)_3$·$6H_2O$	35.29(35.22)	4.17(4.11)	3.73(3.93)	9.22(8.92)	148.5

3.2 SEM and TEM

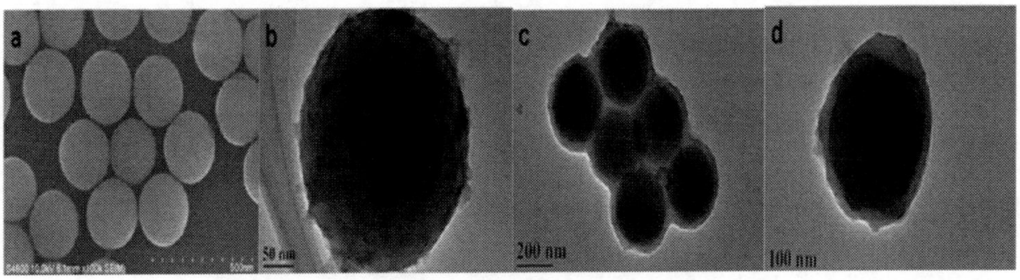

Fig 1. The SEM&TEM images of SiO_2 and SiO_2@Eu(phen-Si)·L, (a) SiO_2(250nm), (b) SiO_2@Eu(phen-Si)·L(20), (c,d) SiO_2@Eu(phen-Si)·L(30)

According to the SEM (Fig.1a), it could be seen that the SiO_2 had regular micro-structure with an average size of 250nm, smooth surface and good dispersity. Fig.1b-d showed the morphologies of core-shell structure SiO_2@Eu(phen-Si)·L with different thickness. It showed that the size of SiO_2 kernel was 250nm, the thickness of the coating layer was 20 and 30nm, which can be expressed as SiO_2@Eu(phen-Si)·L(20), SiO_2@Eu(phen-Si)·L(30). The ternary europium complex had been coated on SiO_2 cores, the SiO_2@Eu(phen-Si)·L core-shell structure had been formed successfully.

3.4 Infrared spectra

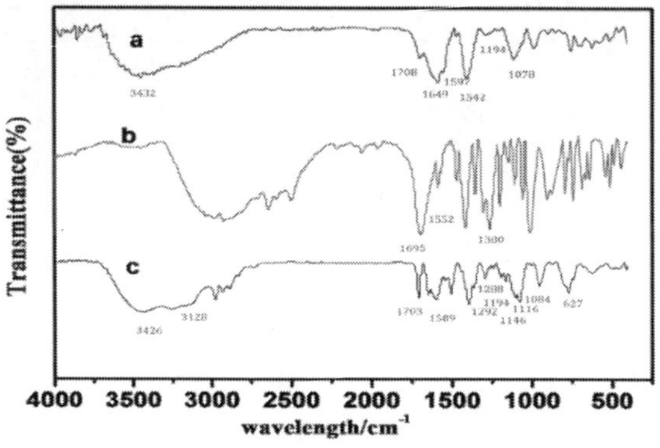

Fig 2. The IR spectra of (a) Phen-Si, (b)(COOH)$C_6H_4SOCH_2COOH$, (c)Eu$(H_2L)_2$·(phen-Si)·$(ClO_4)_3$·$6H_2O$

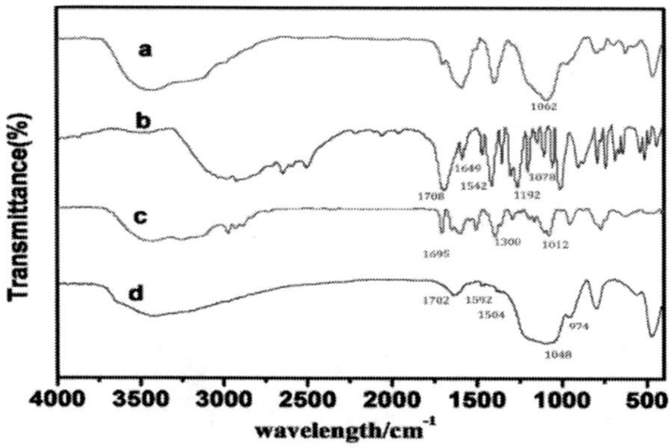

Fig 3. The IR spectra of (a) SiO_2, (b) phen-Si (c)H_2L, (d) SiO_2@Eu(Phen-Si)·$L_{(20)}$

The IR of the two ligands and the Eu(Phen-Si)·$(H_2L)_2$·$(ClO_4)_3$·$6H_2O$ ternary complex could be seen in Fig.2a-c. In the phen-Si IR spectrum (Fig.2a), the νC=O and ν-CONH- appeared at 1708cm^{-1}, 1694cm^{-1} and 1542 cm^{-1} respectively. After coordinating with Eu^{3+}, they shifted to lower wave number which appeared at 1703 cm^{-1}, 1589cm^{-1} and 1552cm^{-1}(Fig.2c). It suggested that europium ion was bonded with phen-Si through the double nitrogen atoms of phen. Similarly, in the sulfoxide IR spectrum (Fig.2b), the νS=O, νC=O(COOH) and νC-OH(COOH) shifted from 1012cm^{-1}, 1695cm^{-1} and 1300 cm^{-1} to 989 cm^{-1}, 1589cm^{-1} and 1292cm^{-1} respectively. Which suggested that rare earth ion bonded with H_2L through the COOH group [14]. When the perchlorate group wasn't coordinated, it was Td symmetry. There were two absorptions bands of 1084~1092 cm^{-1} and 621~629 cm^{-1}. When ClO_4^- was coordinated, it showed C_{2v} symmetry and there were five absorptions [15]. In the IR spectra of the complexes, five absorptions could be seen clearly at 1145, 1116, 1080, 920 and 627cm^{-1}. Combined with the molar conductivity, it could be induced that the one perchlorate group in the complex was bonded with the Eu(III) ion by the oxygen atoms, and the coordination was bidentate.

Fig.3a-d was the IR of the SiO_2, phen-Si, H_2L and the SiO_2@Eu(phen-Si)·$L_{(20)}$. The νC=O, ν-CONH- shifted from 1708cm^{-1}, 1694cm^{-1} and 1542 cm^{-1}to 1702 cm^{-1}, 1592cm^{-1} and 1504cm^{-1} respectively. It indicated that europium ion was bonded with phen-Si through the double nitrogen atoms of phen.

Similarly, in Fig.3c-d, the νS=O shifted from1013 cm^{-1} to 974cm^{-1}, which suggested that europium ion bonded with oxygen atom in sulfinyl group. These results suggest that the complex were formed in the shell layer of core-shell composite.

3.5 Luminescence properties

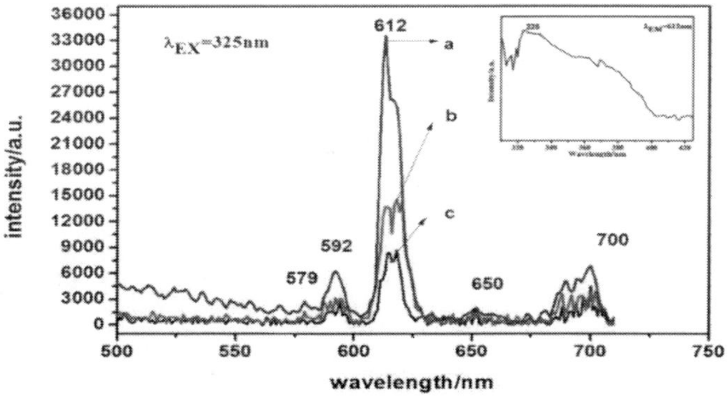

Fig 4. Fluorescent emission spectrum of (a)SiO_2@Eu(Phen-Si)·$L_{(30)}$, (b)SiO_2@Eu(phen-Si)·$L_{(20)}$, (c)Eu(phen-Si)·$(H_2L)_2$·$(ClO_4)_3$·$6H_2O$ Insert: Fluorescent excitation spectrum of Eu(phen-Si)·$(H_2L)_2$·$(ClO_4)_3$·$6H_2O$

The excitation and emission spectrum of the europium complex and the core-shell structure composites were measured in solid state at room temperature (Fig.4). The excitation spectrum of the products was obtained by monitoring their emissions at 612nm. The maximum peak was at 325nm (Fig.4 Insert). There were five emission peaks in the emission spectrum(Fig.4a-c), which were attributed to the characteristic $^5D_0$- ^7F_J (J = 0, 1, 2, 3, 4) transitions of Eu(III) at 579, 592, 612, 650 and 700nm, and the $^5D_0$-$^7F_2$ transition of Eu (III) was the strongest at 612nm. The emission intensity of the Eu(III) ion was 8220 a.u. in europium complex (Fig.4c). However, under the same condition, the emission intensity of SiO_2@Eu(Phen-Si)·$L_{(20)}$ and SiO_2@Eu(Phen-Si)·$L_{(20)}$ core-shell structure composites were 14341 and 32803 a.u. respectively. It was 1.79, 3.99 times greater of the ternary europium complex, respectively. The fluorescence of the core-shell composite was more enhanced than ternary europium complex. On the other hand, according to the Fig.4, the thickness of the coating layer could affect the emission intensity of composites. It meant that when the size of SiO_2 core was fixed, with the increase of the emission intensity was increased with the increase of the thickness of the coating layer.

3.6 The fluorescence decay curve

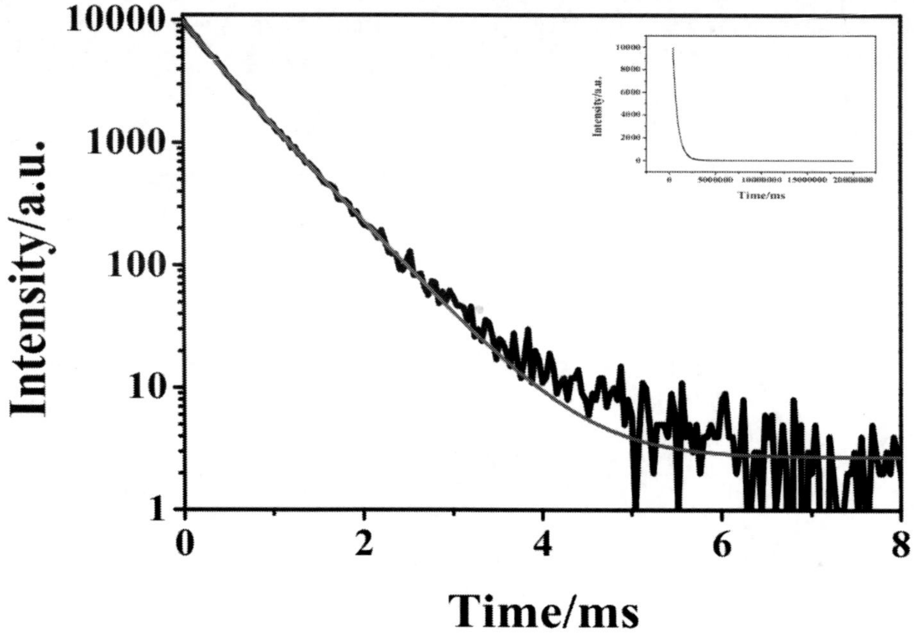

Fig 5. Fit Curves of Eu(phen-Si)·$(H_2L)_2$·$(ClO_4)_3$·$6H_2O$ Insert: Decay Curves of Eu(phen-Si)·$(H_2L)_2$·$(ClO_4)_3$·$6H_2O$

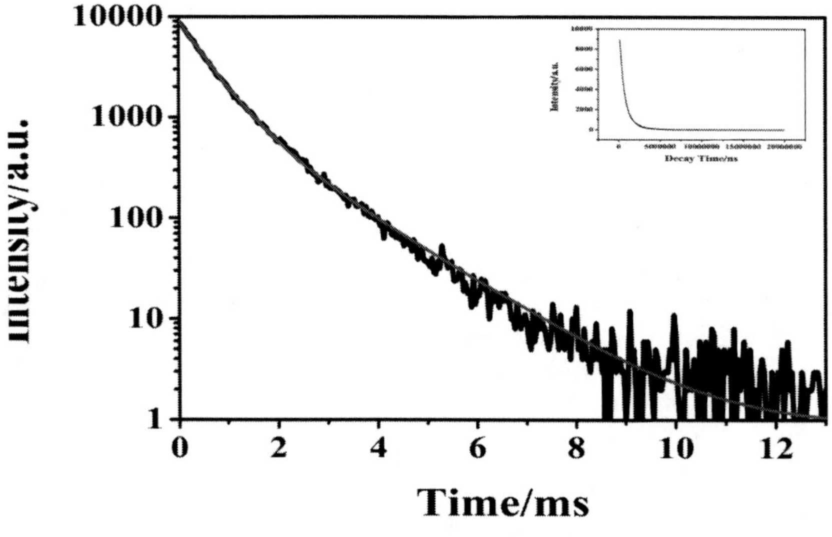

Fig 6. Fit Curves of SiO_2@Eu(Phen-Si)$L_{(30)}$ Insert: Decay Curves of SiO_2@Eu(Phen-Si)$L_{(30)}$

The fluorescence decay curve and double exponential fit curve of the europium ternary complex and SiO_2@Eu(Phen-Si)$L_{(30)}$ core-shell composites was shown in the Fig.5,6. The fluorescence lifetime value of the europium ternary complex and SiO_2@Eu(Phen-Si)$L_{(30)}$ core-shell composites were calculated by the double exponential mode. The lifetime of the Eu(III) ion could be obtained from the Eq. (1),

$$I_{(t)} = I_0 + A_1 \exp\left(-t_1 / \tau_1\right) + A_2 \exp\left(-t_2 / \tau_2\right) \quad (1)$$

$$< \tau > = (A_1\tau_1^2 + A_2\tau_2^2)/(A_1\tau_1 + A_2\tau_2) \quad (2)$$

Where I_t was the fluorescence intensity varying with time t, and τ_1 and τ_2 were lifetime. The corresponding data were calculated. As the result, the lifetime of the europium ternary complex was 527.70 μs, but the lifetime of SiO_2@Eu(Phen-Si)$L_{(30)}$ core-shell composites was 680.00 μs. The results were consistent with corresponding variation tendency of emission intensity.

4 The Conclusion

The novel complex of ternary europium complex Eu(Phen-Si)·$(H_2L)_2$·$(ClO_4)_3$·$6H_2O$, two core-shell structure composites SiO_2@Eu(phen-Si)·$L_{(20)}$ and SiO_2@Eu(phen-Si)·$L_{(30)}$ were prepared. Core-shell structure composites exhibited superior fluorescence properties than that of the ternary europium complex. The emission intensity of the two core-shell composites was 1.79, 3.99 times than the ternary europium complex, respectively. The fluorescence lifetime of the core-shell structure composites was longer than the ternary europium complex. Because rare earth organic complexes were used as coating layer, thickness is nanometer order of magnitude, so the dosage of rare earth organic complexes can be decreased. As a result, the cost of materials will be greatly reduced. The complex and the core-shell structure composite had excellent luminescent properties.

The SiO_2@Eu(phen-Si)·L core-shell structure nano composites had application potential as red luminescent materials and derived application. So this prepared approach provides an applicable new choice for future novel luminescent materials.

This work was supported by the Major projects of Natural Science Foundations of Inner Mongolia Science Foundation (2015ZD01) and the Natural Science Foundations of Inner Mongolia Science Foundation (2015MS0502).

References

1. H.J. Feng, Y. Che, F.Q. Tang, J. Ren. Mater. Lett. **60,** 737–740 (2006).
2. S.C. Ho, D.W. Lee. J. Colloid Interface Sci. **355**, 70-75 (2011).
3. S.X Li, D.M Tang, Z.F Tian, X.J Liu. J. Mater. Chem. C. **5**, 1042–1051 (2017).
4. B.E. William, P.M. Cinta. Opti. Mater. **34**, 355-359 (2011).
5. G.X. Liu, G.Y. Hong. J. Solid State Chem. **178**, 1647-1651 (2005).
6. V. Yu, T.I. Yermolayeva. Radiat. Meas. **46**, 551-554 (2011).
7. D. Sofia, M. Moritz. Langmuir. **27**, 14025–140 (2011) 32.
8. Z.Y Yan, B. Yan. Inorg Chem. Commun. **47**, 96-98 (2014).
9. B. Yan, C. Wang. Inorg Chem. Commun. **14**, 1494-1497 (2011).
10. E. Koft, F.H. Case. J. Org Chem. **27**, 865-866 (1962).
11. E. Meggers, D. Kusch , B. Giese. Helv. Chim. Acta. **80**, 640 (1997).
12. G.M. Kloster, S.P. Watton. Inorg Chim. Acta. **297**, 156 (2000).
13. W. Geary. Coordin. Chem. Rev. **7**, 81-121 (1971).
14. A.Balam, M.L.P. Reddy, M. Jayakanna. J. Phys. Chem. B. **113**, 14128 (2009).
15. M.R. Rosenthal. J. Chem. Educ. **50**, 331-335 (1973).

Comparative Study on Wettability of Typical Plant Leaves and Biomimetic Preparation of Superhydrophobic Surface of Aluminum Alloy

Yanhong Sun[1], *Xinxin* Gao[1], *Jinnong* Li[1], and *Ying* Chen[1*]

[1]Jilin Engineering Normal University, College of Mechanical Engineering, 130052 Changchun, China

Abstract. The water droplets spread observation, contact angle measurement and SEM analysis, data processing and other means, comparative study of different states of the lotus leaf, rice leaf and nasturtium leaf are carried out. The results show that their surface contact angle of three kinds of plant leaves are all over 140 ° , they have hydrophobic or superhydrophobic properties; dehydration degree of leaves has little effect on the wettability, will not change its hydrophobic property; the same point of wettability is that they have micro nano composite microstructure. The difference is, the wettability of the lotus and nasturtium is isotropic, while the rice is anisotropic, that is, in a direction parallel to the edge of hydrophobic than perpendicular to the blade edge enhancement. Based on the theoretical research, the hydrophobic surface is copied to the surface of aluminum alloy. The superhydrophobic surface of aluminum alloy matrix was successfully prepared by chemical etching combined with surface fluorination. The super hydrophobic aluminum alloy surface has micron-nano scale double roughness structure; its surface contact angle reached 156° .

1 Introduction

After billions of years of evolution, natural creatures have acquired near perfect structure and strong environmental adaptability [1]. When the water drops on the lotus leaf surface, not spreads out, but close to the perfect spherical rolling on the lotus leaf surface, falling free, "Pearls falling into a jade plate" fantastic, as shown in Fig. 1. The water drops from the lotus leaves can carry away the dust, so the lotus leaves have "self-cleaning" characteristics. In 1977, William Barthlott first named the self-cleaning phenomenon known as the "lotus leaf effect" (Lotus- effect)" [2]. It contains profound scientific principles—"Infiltrative" [3,4]. It is governed by the laws of science. If the contact angle of water drop on solid surface is greater than 150 ° , it is called super hydrophobic surface [5], lotus leaves are super hydrophobic.

Although experts and scholars have made many achievements on superhydrophobic surface of plants. But, not to carry out a comparative study on the hydrophobic mechanism of typical plants, research on the causes of the difference in wetting function of typical

* Corresponding author: 823068684@qq.com

plants. The superhydrophobic mechanism and the similarities and differences of lotus leaf, rice and nasturtium are not reported.

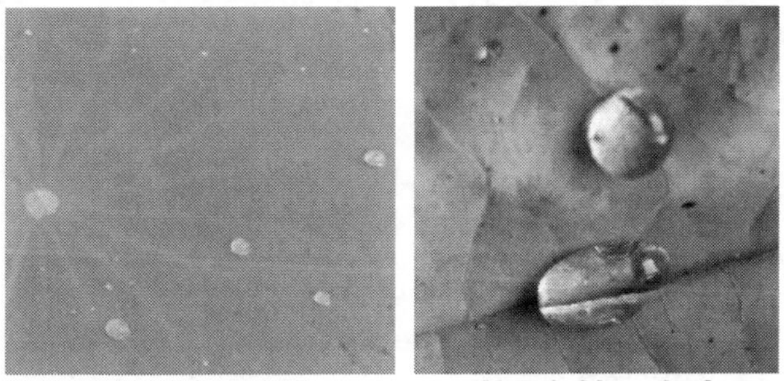

(a)Fresh lotus leaf (b) Dried lotus leaf

Fig. 1. Drops of water on lotus leaves

2 Equipment

Using S-300N (Japanese manufacturing HITACHI) SEM microstructure and morphology analysis of plant leaf surface observation; Application of DSA100 surface analysis system (German manufacturing company of Kruss) and water contact angle measurement of the sample surface, water volume is 5 L.

3 Wettability contrast experiment for typical plants of different dehydrated leaves

Fresh plant leaves are picked up after the source of moisture is cut off. As time goes on, the surface moisture decreases gradually. The influence of dehydration degree on the wettability of leaves was investigated. Through the direct observation of the distribution of water droplets on the blade, contact angle measurement and microscopic analysis, the wettability of the three samples was compared.

3.1 Lotus leaf

Lotus is a large perennial herb. It is long rhizome, leaves come directly from the root, long stalked, spiny, blade peltate, projecting above water[1].

As shown Fig. 1, the difference of surface wettability of between the fresh and the dried lotus leaves was observed. For the fresh leaf, the water droplets together, forming a large drop of water, and can roll freely, showed a strong hydrophobic function; For the dried leaf, little drops by two and three everywhere, the hydrophobic property is not inferior to the fresh leaves. Therefore, lotus leaves showed hydrophobic characteristics in two kinds of state. The contact angle of lotus leaves is shown in Table 1.

Table 1. Measurement results of contact angles of lotus leaves

Types and locations of measurements	**Contact angle[°]**
Fresh lotus leaf	151±3
Dried lotus leaf	155±3

The Fig. 2 shows the morphology of lotus leaf surface. Its structure is composed of a large number of evenly distributed micro hemispherical papillae of about 10 ~ 15 μm diameter, 2-5 μm height, 10 ~ 20 μm distance, which are covered with 200 ~ 500nm nanometer wax villi on the mastoid. Water droplets in each direction is free rolling on the surface, it is easy to assemble a big drop, so super hydrophobic characteristic of lotus leaf surface with isotropic and strong catchment characteristics.

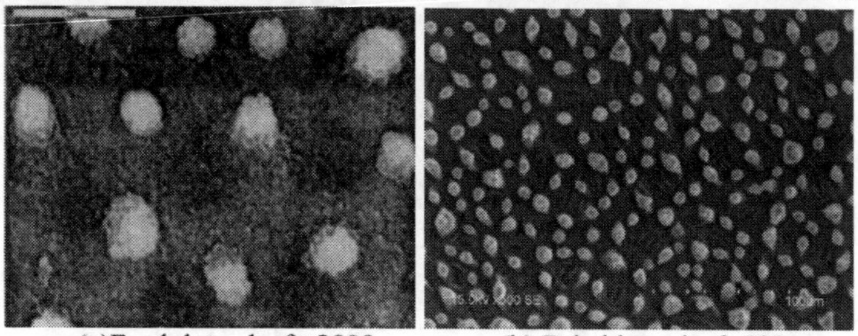

(a)Fresh lotus leaf ×2000 (b) Dried lotus leaf×500

Fig. 2. Lotus leaf SEM contrast photos

3.2 Rice leaf

The spreading of water droplets on the leaf surface of rice is shown in Fig. 3 Whether fresh or dried leaves, the water droplets are round and show better hydrophobic properties. Through the observation of rice leaf surface SEM diagram (refer with: Fig. 3), rice leaf surface micro-nano hierarchical structure is similar to the lotus leaf, by the diameter of micron 5~8 micro convex body and its diameter is about 20~50nm acicular wax granules composed of two element composite structure, but in the rice surface, micron mastoid in parallel with the leaf edge direction (arrow), orderly arranged in the vertical direction of leaf edge are arranged disorderly. The angle between rice leaves and water was 142 ° , showing good hydrophobic and super hydrophobic characteristics. Rice leaves above the water droplets can roll along the direction parallel to the edges of the leaves of freedom, the direction of the rolling angle (3 ~5°) than the rolling edge of the vertical leaf angle (9 ~15°), expressed as hydrophobic properties anisotropy [6].

(a)Fresh leaf (b)Dried leaf (c)SEM photo

Fig. 3. Wetting of water droplets on rice leaf samples

3.3 Nasturtium

Nasturtium is the rare wild plants of ranunculaceae. Its leaves are made up of many small hills high in 10~15um, as shown in Fig. 4(c). The hills are distributed on the surface of 100~150nm nanometer scale clusters, 50nm width nano column pile up. Thus, micro-nano structure ordered. The contact angle of nasturtium leaf is 145°, super hydrophobic.

With the increase of dehydration degree on the blade surface, the static contact angle has also changed, and the experimental results are shown in Fig. 4(d). With distilled water as a test fluid, the changing is shown in Fig. 5.

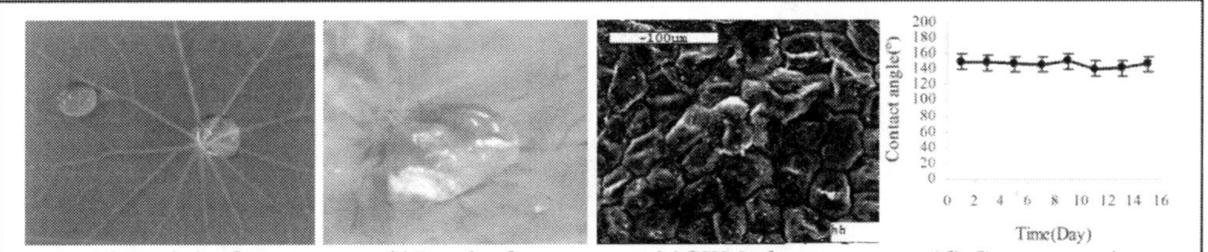

(a)Fresh leaf (b)Dry leaf (c)SEM photo (d) Contact angle

Fig. 4. Nasturtium leaf wetting condition

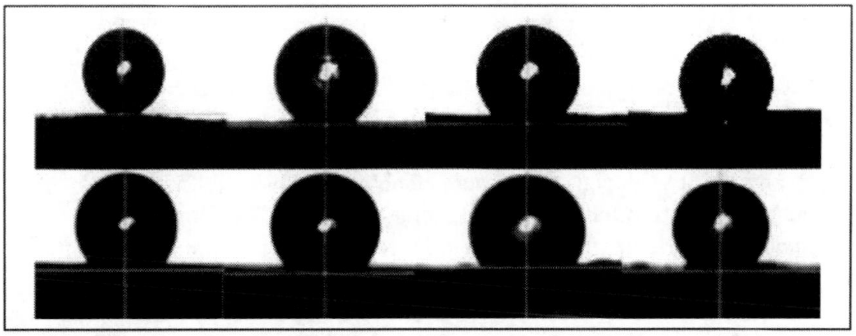

Fig. 5. Contact angle changing during drying

3.4 Wettability comparison and analysis

Barthlott and Neinhuist [3] have revealed for the first time that the superhydrophobic nature of lotus leaf surface is the result of the combination of micro- nano structure on its surface. On the basis of this research, Jiang Lei pointed out that the roughness caused by the micro-nano structure of the lotus leaf surface is the fundamental reason for the super hydrophobic and self-cleaning of lotus leaves. From the water droplets in the leaf surface spreading, water contact angle measurements and SEM observation etc. we found that several plant leaf scales have good hydrophobic or super hydrophobic properties, they have micro-nano composite microstructure that is because they are so with micro-nano morphology, which is the main reason of hydrophobic or super hydrophobic properties of typical plant leaves.

The leaves of these three kinds of plants have different wettability. The lotus leaves and nasturtium leaves are isotropic, while the leaves of rice are anisotropic.

4 Study on biomimetic application of aluminum alloy materials

It is found that the micro-nano composite structure has important reference value for the preparation of biomimetic superhydrophobic surface by studying the leaf table of superhydrophobic plants[7,8]. By combining chemical etching with surface modification, the micro - nano composite structure on the surface of aluminum alloy has been successfully produced, which makes it appear super hydrophobic[9,10].

4.1 Surface modification of aluminum alloy by chemical etching

Aluminum Alloy is divided into rectangular pieces of 2 × 2.5(cm), polished with fine sandpaper, make it look very smooth, followed by acetone soaking aluminum decontamination, deionized water ultrasonic cleaning, and then immersed into 10min hydrofluoric acid solution of 3.75mol/L to remove the surface oxide layer.

The chemical etching solution was prepared with HC1, HNO and HF with volume ratio of 15:5:1, and the aluminum sheet was soaked in corrosive solution at room temperature for chemical etching. When corrosion, gently clamp the surface of aluminum sheet with tweezers, turn over in the corrosive liquid, and make a large amount of gas overflow. After the reaction is complete, the aluminum sheet is quickly flushed with a large amount of water, and the surface is scrubbed with absorbent cotton. After drying, the rough micro / nano structure with two different scales is prepared[11].

4.2 Modification of micro nano structured surface with Fas

It is not enough to construct micro / nano composite structures on the surface of metal substrate. After using the low surface energy material to modify the micro nano structured metal surface, it has both micro nano structure and very low surface energy.

At room temperature, Fas and water are mixed at a volume ratio of l:49 to form a 2% intermediate solution of Fas. Then, the PH value of the intermediate solution is adjusted between 3 and 3.8 by glacial acetic acid. The amount of adding Fas to dilute ethanol, arranged in 1% (WT) fluoroalkyl alcohol solution, stirring for about 5 hours at the 50 r/min rate in a magnetic stirrer. After forming a homogeneous and transparent solution, the Fas solution is placed to be used[12].

After the chemical etching, the surface of the aluminum sheet has hydrophobic characteristics, but the surface is hydroxylated. The aluminum fluoride soaked in Fas solution for 30 min, the hydroxylated surface reacts with the silane hydrolysis group, forming a layer of nano FAS-SAM, the final sample placed one or two days to be dried, combined with Fas nonstick film is completely formed.

4.3 Test results

The Fig.6 is the SEM photos of aluminum alloy surface etched by strong acid solution for 12 min after being zoomed 3000 magnification, contact angles photos of water droplets on superhydrophobic aluminum alloy surface are also visible. On the surface of the etched aluminum alloy, the labyrinth structure are formed by consists of many rectangular bosses and pits. A micronscale smaller boss is also distributed on the nanoscale boss. There are many nano scale bump on micro scale convex platform, and these micro and nano convex layers are stacked into ladder like distribution. This shows that the super hydrophobic aluminum alloy surface has micron and nano scale double roughness structure. The contact angle is 156°. It shows that it has very good super hydrophobic characteristics.

In addition, the static contact angle data of the superhydrophobic aluminum alloy surface were not changed obviously after 6 months in the air. This shows that the surface of the superhydrophobic aluminum alloy prepared in the experiment has good time stability.

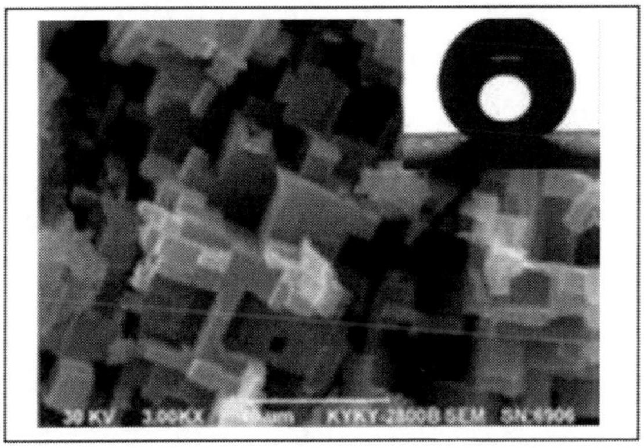

Fig. 6. EM photos of etched aluminum alloy surface

5 Conclusions

Through the measurement of static contact angle, scanning and data processing, the positive and negative wettability of lotus leaves was studied. The results showed that the degree of dehydration had little influence on wettability, and hydrophobicity did not change.

A comparative study of the hydrophobic mechanism shows that the hydrophobic mechanism of lotus, nasturtium and rice leaves is because of their micro-nano composite structure. Both the lotus leaves and nasturtium leaves have isotropic wetting properties, while the leaves of rice are anisotropic, i.e., in parallel with the blade edge, the hydrophobicity is stronger than that perpendicular to the blade edge.

Modeled on the surface of lotus leaves and nasturtium super hydrophobic and hydrophobic, the surface of aluminum alloy is chemical etched and low surface energy material modified to form micro-nano ladder rough structure of bosses and pits. Its contact angle is 156°showing good super hydrophobic characteristics.

References

1. L. Feng, H. S. Li, L. Jiang. Adv. Mater, 24, **14** (2002), p1857
2. C. Neinhuis, W. Barthlott. Annals of Botany, 79,**6** (1977) , p.667
3. W. Barthlott, C. Neinhuis. Planta , 202, **1** (1997)1, p.1
4. A. B. Cassie, S. Baxter. Trans Faraday Soc, 40, **8** (1944), p546
5. A. Lafurna, D. Quere. Nature Mater, 7, **2** (2003)2, p.457
6. X. F. Zhao, M. Q. Yang, L. Zhang. Chemical Industry and Engineering Process, 35, **9** (2016), p.2812(In Chinese)
7. A. Terray, J. Oakey, D. W. Marr. Science, 5574, **296** (2002), p.1841
8. A. D. Tserepi, M. E. Vlachopoulou, E. Gogolides. Nanotechnology, 17, **15** (2006), p.3977
9. M. H. Sun, C. X. Luo, L. P. Xu. Langmuir, 19, **21** (2005) 1, p 8978(In Chinese)
10. J.L. Song. Study on surface preparation and application of extreme wettability of engineering metal materials (Ph.D., Dalian University of Technology, China 2015), p.42 (In Chinese)
11. A . Nakajima, D. Fujishima, K. Hashimoto. Adv Mater, 16, **11** (1999)1, p.1365
12. R. N. Wenzel. Industrial and Engineering Chemistry, 28, **8** (1936), p.988

AUTHOR INDEX

EDP Sciences – Web of Conferences
17, Avenue du Hoggar
Parc d'Activité de Courtabouf
BP 112
F-91944 Les Ulis Cedex A
France

ISBN 978-1-5108-5579-3

International Workshop on Transportation and Supply Chain Engineering (IWTSCE'16)

MATEC Web of Conferences Volume 105 (2017)

Rabat, Morocco
29 - 30 November 2016

edp sciences